TOP GUN

TOP GUN

Christopher Chant

Quantum
Books

A QUANTUM BOOK

This book is produced by
Quantum Publishing Ltd.
6 Blundell Street
London N7 9BH

Copyright ©MCMXCII
Quintet Publishing Ltd.

This edition printed 2002

All rights reserved.
This book is protected by copyright. No part of it may be
reproduced, stored in a retrieval system, or transmitted in
any form or by any means, without the prior permission
in writing of the Publisher, nor be otherwise circulated in
any form of binding or cover other than that in which it
is published and without a similar condition including
this condition being imposed on the subsequent
publisher.

ISBN 1-86160-306-1

QUMTOF

Typeset in Great Britain by
Central Southern Typesetters, Eastbourne
Manufactured by J Film Process Singapore (Pte) Ltd
Printed in Singapore by
Star Standard Industries (Pte) Ltd.

The material in this book previously appeared in
Aviation Recordbreakers and
Aircraft Prototypes

CONTENTS

BASIC DESIGN AND STRUCTURE 7

POWERPLANT 73

PROTOTYPES 99

- AMERICAN BOMBERS 100

- SOVIET BOMBERS 116

- BOMBERS FROM OTHER COUNTRIES 124

- AMERICAN FIGHTERS 128

- SOVIET FIGHTERS 154

- FIGHTERS FROM OTHER COUNTRIES 174

- RESEARCH AIRCRAFT 196

INDEX 206

1
Basic Design and Structure

The late 1980s are seeing dramatic development right across the board of aircraft design. Many of the premises that have been taken for granted for half a century are now being questioned, and a new generation of aircraft is offering completely unforeseen capabilities. Military aircraft are benefiting not in outright performance, which has reached a plateau of what is technically realistic, but rather from features that enhance their tactical capabilities. At the same time, civil aircraft are becoming safer and more economical both to operate and to maintain. The requirements of military and civil aircraft may appear incompatible, but in fact many of the technologies are shared, though in different ways and to different degrees.

The driving force behind the current spate of developments is the military's desire to introduce new generations of combat aircraft. Until the mid-1960s the chief concern of the world's more advanced air arms had been the development and deployment of combat aircraft characterized by extremely high outright performance figures, particularly for speed, rate of climb, service ceiling and range. The first three depended on a clean aerodynamic

---

McDONNELL DOUGLAS F-4F PHANTOM II

TYPE: two-seat air-superiority fighter
WEIGHTS: empty 30,328 lb/13,757 kg; maximum take-off 61,795 lb/28,030 kg
DIMENSIONS: span 38 ft 7½ in/11.77 m; length 63 ft 0 in/19.20 m; height 16 ft 5½ in/5.02 m; wing area 530 sq ft/49.24 m²
POWERPLANT: two 17,900-lb/8,119-kg afterburning thrust General Electric J79-GE-17A turbojets
PERFORMANCE: speed 1,430 mph/2,301 km/h; ceiling 58,750 ft/17,905 m; range 1,424 miles/ 2,290 km
ARMAMENT: one 20-mm Vulcan multi-barrel cannon and 16,000 lb/7,257 kg of disposable stores (including six Sparrow medium-range or four Sparrow and four Sidewinder short-range air-to-air missiles, Maverick air-to-surface missiles, rocket pods and a wide assortment of free-fall bombs) on four special missiles stations plus one underfuselage and four underwing hardpoints

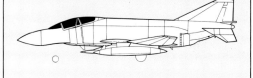

OPPOSITE The greatest Western combat aircraft of the period immediately after World War II, the McDonnell Douglas Phantom II (here the 5,000th aircraft, an F-4E for Turkey) has wholly distinctive aerodynamics and is still a potent warplane some 30 years into its career.

ABOVE Among the earliest Phantom IIs were this pair of F4H-1Fs for the US Navy, who gradually brought them up to definitive F-4B standard.

LEFT The Phantom II also served as an experimental aeroplane in the form of this F-4CCV, fitted with canard foreplanes for enhanced agility, and retaining the experimental fly-by-wire control system pioneered by this particular aeroplane in earlier trials.

LEFT With a few exceptions modern combat aircraft are too expensive for use in a single role, and sights such as this F-4D are common: the aeroplane carries AIM-7 Sparrow air-to-air missiles and laser-guided 'Paveway' guided bombs for a dual, air combat and ground-attack capability.

design combined with a high power-to-weight ratio, and the last on clean aerodynamic design combined with fuel-economical engines and large fuel capacity.

The result was aircraft such as the McDonnell Douglas F-4 Phantom II multi-role fighter with a speed of about Mach 2.25, an initial rate of climb of 30,000 ft/9,145 m per minute, a service ceiling of 60,000+ ft/18,290 m and a typical tactical radius of 700 miles/1,125 km. The subsequent replacement of turbojets by turbofan engines resulted in a useful increase in tactical range on a given fuel load.

The limiting factor on designs of the 1960s and 1970s was the use of aluminium alloys for the primary structure. Such alloys are available in a vast range of tailored varieties, offer considerable strength at modest weight, and do not present great difficulties in airframe manufacture. However, they do lose strength at high temperatures, a factor which becomes critical at airspeeds of about Mach 2.5. Here aerodynamic heating of the airframe begins to present insuperable problems for an aluminium alloy structure. They can be replaced by steel and titanium, but such materials are expensive both to manufacture and to work, and are therefore limited to use in critical high-temperature areas such as wing leading edges and the vicinity of engine nozzles. The only exceptions are specialized aircraft such as the American Lock-

LOCKHEED SR-71 CREW

Appropriately named Blackbird in view of the colour of its airframe's heat-reducing and radar-absorbent coating, the Lockheed SR-71 is an aircraft whose real performance clearly exceeds the figures set by the type in several world record-breaking flights. These include the records for absolute speed in straight-line flight (2,193.17 mph/3,529.56 km/h, established on July 28, 1976), speed over a 1,000-km/621.1-mile closed circuit (2,092.294 mph/3,367.221 km/h on July 27, 1976) and sustained height in horizontal flight (85,059 ft/ 25,929.03 m on July 28, 1976).

The SR-71 operates on the fringes of the atmosphere and the two-man crew – pilot and rear-seat radar systems operator – dress like astronauts: indeed, the S-1010B full-pressure suit designed for the crew of Lockheed U-2 and SR-71 reconnaissance aircraft was used by the astronauts on the first Space Shuttle flights.

While the aircraft is undergoing its time-consuming pre-flight preparation the crew members undergo a full medical examination by the Physiological Support Division attached to each operational unit. It is important that nitrogen be purged from the body to avoid the possibility of the bends, and for about an hour the two men breathe pure oxygen to achieve this end. The men are also given anti-fatigue drugs as a means of reducing muscle stiffness on long flights, and the PSD team check that the crew's special diet has succeeded in reducing bowel activity.

The S-1010B suit is completely sealed, and until the crew members have boarded the aircraft and plugged themselves into the onboard life-support system they have to carry lightweight air-conditioning units to prevent themselves overheating. And at the end of the mission the crew are examined again by the PSD team after being debriefed by the unit intelligence team.

heed SR-71 Blackbird high-altitude strategic reconnaissance aircraft, which has a primary structure of titanium for Mach 3+ performance, and the Soviet Mikoyan MiG-25 Foxbat high-altitude interceptor, which employs a steel primary structure for similar speeds.

Consequently, pending the development of a new structural medium that was both comparatively inexpensive and straightforward to use, mass-production combat aircraft were generally limited to a top speed of Mach 2.5. It was the Americans who first pushed to such a limit in their constant search for tactical aircraft that would have a performance edge over their Soviet counterparts, allied to a technological edge in factors such as radar, other electronic systems and guided missiles.

LOCKHEED SR-71A BLACKBIRD

TYPE: two-seat strategic reconnaissance aircraft
WEIGHTS: empty 60,000 lb/27,216 kg; maximum take-off 170,000 lb/77,111 kg
DIMENSIONS: span 55 ft 11 in/16.94 m; length 107 ft 5 in/32.74 m; height 18 ft 6 in/5.64 m; wing area about 1,000 sq ft/92.9 m²
POWERPLANT: two 32,500-lb/14,742-kg afterburning thrust Pratt & Whitney JT11D bleed turbojets
PERFORMANCE: speed 2,250 mph/3,620 km/h; ceiling 100,000 ft/30,480 m; range 2,980 miles/4,800 km
ARMAMENT: none

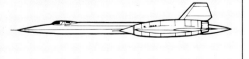

ABOVE Currently the world's ultimate aeroplane, the Lockheed SR-71A strategic reconnaissance platform is highly classified, but holds world records for speed and altitude. The special paint helps to radiate heat and is also radar-absorbent.

MIKOYAN MiG-25 FOXBAT-A

TYPE: single-seat interceptor
WEIGHTS: empty 44,092 lb/20,000 kg; maximum take-off 82,507 lb/37,425 kg
DIMENSIONS: span 45 ft 9 in/13.95 m; length 78 ft 1¾ in/23.82 m; height 20 ft 0¼ in/6.10 m; wing area 611.7 sq ft/56.83 m²
POWERPLANT: two 27,116-lb/12,300-kg afterburning thrust Tumanskii R-31 turbojets
PERFORMANCE: speed 1,849 mph/2,975 km/h; ceiling 80,050 ft/24,400 m; range 1,404 miles/2,260 km
ARMAMENT: four AA-6 Acrid long-range air-to-air missiles on underwing hardpoints

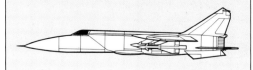

THE VIETNAM EXPERIENCE

The fallacy behind US combat aircraft design was revealed during the Vietnam War, when the US air arms were faced only by the tiny North Vietnamese air force with its small number of obsolescent fighters and large numbers of indifferent surface-to-air missiles, both of Soviet origin. In retrospect it is clear that the surface-to-air missiles posed only a modest threat, though their availability in ever-increasing numbers dictated the development of electronic countermeasures and new tactical gambits. The North Vietnamese fighters also posed only a small threat because of their tiny numbers.

Far more worrying were the fundamental tactical deficiencies revealed when the Americans' latest fighters tackled Soviet-supplied

TOP Successor to the classic MiG-21, the Mikoyan-Gurevich MiG-23 is a capable air-defence and multi-role fighter with modern radar and variable-geometry wings.

ABOVE MIDDLE The MiG-25 has phenomenal performance in a straight line at high altitude, but is less capable against current targets at lower speeds and lower levels. This deficiency is partially remedied in this, the 'Foxbat-E' version.

ABOVE The massive aft-fuselage powerplant installation of the 'Foxbat-E'.

MiG-19 Farmer fighters. The American fighters were designed to use sheer performance to achieve a position where their radars could detect opposing fighters at medium ranges, then engage them with missiles such as the medium-range radar-guided AIM-7 Sparrow or short-range heat-seeking AIM-9 Sidewinder. But the radars of the American fighters proved unreliable and the US rules of engagement dictated that the target must be identified visually before missile launch, ruling out medium-range Sparrow attacks. And at visual ranges the superior tactical capabilities of the Soviet fighters proved a decided embarrassment to the Americans. The MiG-19 is capable of only Mach 1.35, but as the Americans rapidly discovered, any short-range turning engagement almost immediately causes the combatants to lose energy. The resulting drop in speeds toward Mach 0.8 or so meant the better turning rate of the lighter fighter became decisive.

At relatively low speeds and altitudes the Soviet fighters had a decided edge in the real tactical requirements of acceleration, rate of climb and rate of turn. Consequently, in visual combat they were able to outmanoeuvre the larger and heavier American fighters and bring their devastating cannon armament to bear. To make matters worse, the Americans had decided that guns were obsolete, and fighters such as the early Phantoms had been designed without them.

As a short-term expedient podded 20-mm cannon could be carried under the wings or fuselage, but these lacked the installational rigidity of built-in guns and accordingly were less accurate than the Mig-19's three 30-mm weapons.

BELOW Long in the tooth but still a useful weapon and still under further development, the AIM-7 Sparrow is a medium-range missile which homes onto radar energy 'bounced' off the target by the launch fighter's radar.

NAVAL WEAPONS CENTER AIM-9 SIDEWINDER

Designed originally by the Naval Ordnance Test Station (now the Naval Weapons Center) at China Lake, California, and subsequently produced in improved versions by Ford Aerospace and Raytheon, the AIM-9 Sidewinder is the West's most important short-range air-to-air missile. Its development, which encapsulates the whole history of infra-red AAM guidance, began in 1949 using a body of only 5-in/127-mm diameter, and full-scale production was entrusted to Philco Ford (now Ford Aerospace) in 1951.

The first XAAN-N-7 guided prototype was fired in September 1953, and the SW-1 initial production variant reached operational capability in 1956. This model became the AIM-9B in the US forces' tri-service rationalization of designations during 1962. Production reached 80,900 units, and the 155-lb/70.4-kg missile could reach a range of 2 miles/3.2 km with its 20-second mission endurance. The lead sulphide seeker was uncooled, and the claimed 70% single-shot kill probability was attainable only under perfect conditions in a rear-hemisphere attack giving the seeker an unobstructed view of the target's jetpipe.

Subsequent development has resulted in many versions, including the limited-production AIM-9C with semi-active radar homing. The most important developments have concerned: seeker sensitivity for all-aspect engagement capability; greater agility to give dogfighting performance against a manoeuvring target; a more destructive warhead; greater speed; longer range; and a smokeless motor to reduce the giveaway plume that characterized all models up to the AIM-9L.

Notable versions were:
- AIM-9D with a nitrogen-cooled seeker, greater speed and range, and a new 22.4-lb/10.2-kg continuous-rod annular blast/fragmentation warhead in place of the original 10-lb/4.54-kg blast type;
- AIM-9E, with a wide-angle seeker;
- AIM-9G, with off-boresight lock-on capability;
- AIM-9H with solid-state electronics;
- AIM-9L with all-aspect seeker capability and greater agility;
- AIM-9M, with greater resistance to countermeasures;
- and AIM-9N with considerably enhanced agility.

The AIM-9R is modelled on the AIM-9N but is more reliable; it weighs 172 lb/78 kg and can reach a range of 11 miles/17.7 km during its 60-second flight-time. The AIM-9 is due to be replaced in the 1990s by the European AIM-132 ASRAAM, but development is continuing.

The Americans moved rapidly to mitigate these shortcomings by enhancing existing aircraft. The F-4E version of the Phantom was produced with a revised nose housing a multi-barrel 20-mm cannon. Subsequently, to improve dogfight agility the outer wing panels were fitted with leading-edge slats. The latter replaced the blown leading-edge flaps incorporated into the original design to reduce landing speeds on carrier decks.

A NEW GENERATION

In the longer term, of course, the Americans saw that they needed a new generation of combat aircraft. The myth of outright superiority in performance had been exploded over Vietnam, and the US services now appreciated what the Soviets had known all along: performance at high altitude was all very well, but surface-to-air missiles had forced operating altitudes down to levels where Mach 2 speeds were not readily attainable and where acceleration, both straight-line and turning, was all-important. Bombers had already been re-cast in the low-level role, in the Boeing B-52 Stratofortress high-altitude strategic bomber being strengthened for its new operating regime and the extraordinary North American B-70 Valkyrie high-altitude Mach-3 bomber – against whose threat the Soviets produced the MiG-25 Foxbat – being cancelled. Now it was the turn of the fighters.

In common with all other aircraft types, the fighter is designed to embody a blend of characteristics optimized for its specific role, whether it be interception, air superiority or air combat. Followed by the Europeans, the Americans now decided that the blend should be revised for greater agility at lower altitudes. Good performance was still needed, but it was clear that the maximization of performance to the detriment of other elements had unbalanced the whole blend.

The new generation of fighters is epitomized by the General Dynamics F-16 Fighting Falcon, winner of a US Air Force competition for a lightweight fighter. The competition was initially seen as a technology demonstration that would evaluate aerodynamic concepts. But by 1975 it was clear that it had produced two prototypes with exceptional combat potential (the General Dynamics YF-16 and Northrop YF-17) and the competition was revised to

LEFT The modern combat aeroplane is increasingly a flying sensor and computing platform to which can be attached an increasing variety of weapons. This is an F-4E armed with an AGM-65 Maverick, a small but highly impressive air-to-surface missile available in variants with different guidance and warhead packages.

RIGHT This F-4E is seen en route to a target in Vietnam during the South-East Asia War, and carries two triplets of 'iron' bombs with nose probes to ensure detonation above the ground.

LEFT This Phantom II's impressive external load includes one drop tank, one electronic countermeasures pod, three Sparrow air-to-air missiles, and two AGM-45 Shrike anti-radar missiles.

ABOVE The North American B-70 Valkyrie supersonic strategic bomber was overtaken by the pace of tactical developments and never entered service, but was in every way a remarkable aircraft of superb performance, and was so highly regarded by the Soviets that they developed the MiG-25 expressly to counter the threat it posed.

ABOVE RIGHT After its termination as a bomber prototype, the B-70 was operated by the US Air Force and NASA as an experimental aeroplane with a number of aerodynamic and structural spin-offs for emerging programmes.

produce a multi-role combat aircraft for the USAF's Tactical Air Command. The rationale behind the F-16 was the desire for a lighter, and therefore cheaper, partner for the massive McDonnell Douglas F-15 Eagle air-superiority fighter. The F-16 entered US service in 1979, and has been adopted subsequently by many of the USA's allies. Meanwhile, the loser in the competition, the YF-17, became the basis for the McDonnell Douglas F/A-18 Hornet dual-role fighter and attack aircraft. The Hornet bears some aerodynamic similarities to the F-16, though it has larger wing/fuselage strakes, twin vertical tail surfaces and twin engines.

CONTROL-CONFIGURED VEHICLES

The YF-16 had been designed from the start as a control-configured vehicle (CCV) of modest performance. A conventional aeroplane is naturally stable, with its centre of gravity located forward of its centre of lift, and requires balance by constant down elevator, resulting in down-load at the tail which is increased considerably in supersonic flight. A CCV, on the other hand, is naturally unstable, with its centre of gravity well aft of its centre of lift: control is effected by powerful aerodynamic surfaces working in association with a computer-directed automatic flight-control system, resulting in up-load at the tail in subsonic flight and only modest down-load in supersonic flight.

The key to CCV design is relaxed static stability, which means that left to itself the airframe would turn end-over-end immediately after takeoff. However, the automatic flight-control system uses its air data system, accelerometers and rate gyros to sense any such tendency immediately, and can then apply corrective control-surface deflections at a rate of more than 100 per second.

Given an automatic flight-control system the weight distribution of the CCV aeroplane can be optimized for agility rather than stability, the computers ensuring that safe flight is maintained and that the pilot's control inputs – on the F-16 these are effected through a small side-stick rather than a centrally-mounted control column – are immediately translated into the optimum control movements. Such a system offers unprecedented levels of agility within the structural limits of the airframe, as well as reducing gust response and structural stress. The last two factors allow a slightly lighter airframe to be used and extend airframe life.

McDONNELL DOUGLAS F-15A EAGLE

TYPE: single-seat air-superiority and attack fighter
WEIGHTS: empty 27,000 lb/12,247 kg; maximum take-off 68,000 lb/30,845 kg
DIMENSIONS: span 42 ft 9¾ in/13.05 m; length 63 ft 9 in/19.43 m; height 18 ft 5½ in/5.63 m; wing area 608 sq ft/56.5 m²
POWERPLANT: two 23,950-lb/10,864-kg afterburning thrust Pratt & Whitney F10-PW-100 turbofans
PERFORMANCE: speed 1,650+ mph/2,655+ km/h; ceiling 60,000 ft/18,290 m; range 2,878+ miles/4,631 km
ARMAMENT: one 20-mm Vulcan multi-barrel cannon and up to 16,000 lb/7,258 kg of disposable stores (including four Sparrow medium-range and four Sidewinder short-range air-to-air missiles, guided glide bombs, rocket pods and an exceptionally wide assortment of free-fall conventional or cluster bombs, as well as electronic pods of varying types) on four tangential, three underfuselage and two underwing hardpoints

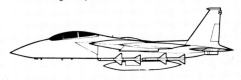

The F-16 is of orthodox configuration, with a single vertical tail surface, mid-set all-moving tailplane and mid-set wings of 40° leading-edge sweep. A closer examination reveals a number of aberrations from the norm, however:
- the tailplane is of the taileron type, the two halves operating in unison for pitch control and differentially for roll control;
- the full-span ailerons are of the flaperon type, again operating differentially for roll control and in unison for increased lift;
- the leading edges are hinged to allow the automatic flight-control system to schedule them as automatic leading-edge flaps and, in concert with the 'flaperons', provide a measure of variable camber to the wing;

McDONNELL DOUGLAS F-15E STRIKE EAGLE

The McDonnell Douglas F-15 Eagle, planned as an air-superiority fighter to succeed the McDonnell Douglas F-4 Phantom II and the older Convair F-106 Delta Dart, began to enter service in November 1974. The type was an immediate success, with a thrust/weight ratio exceeding unity and consequent excellent climb and high-altitude performance, and proved an admirable attack fighter into the bargain.

The original single-seat F-15A and two-seat F-15B variants – the latter being a combat-capable proficiency trainer – have matured as the F-15C and F-15D. The later versions have upgraded electronics and the ability to carry FAST (Fuel And Sensor Tactical) packs, which attach conformally along the outside of the engine trunks under the wings to increase fuel capacity and electronic capability at minimal cost in weight and drag. The FAST packs are also stressed for the tangential carriage of additional weapons.

The F-15D formed the basis for the F-15E Strike Eagle, an attack-optimized version under development for the US Air force. The first example flew in 1987, and the type offers superb capabilities well suited to the operational requirements of the European theatre. The Strike Eagle was first suggested as the private-venture F-15 Enhanced Eagle, whose two-man crew concept has been used in the Strike Eagle.

The F-15E's main radar, the highly capable Hughes APG-70, is supported by an extensive suite of defensive electronics. The pilot has a wide-angle head-up display, head-down displays and a moving map display, while the rear-seater has no fewer than four head-down displays for weapon management and threat monitoring. The F-15E can also lift a maximum weapon load of 24,250 lb/11,000 kg, compared with the F-15C's 16,000 lb/7,258 kg, can carry all the latest weapons, and is designed to use the podded LANTIRN (Low-Altitude Navigation and Targeting Infra-Red for Night) system for all-weather missions at very low altitudes.

The engine bay is of modular design able to accept either the Pratt & Whitney F100 or the General Electric F110, afterburning turbofans which will eventually deliver a thrust of more than 30,000 lb/13,608 kg. This level of power will allow the F-15E to take off at a weight of 81,000 lb/36,741 kg compared with the F-15C's 68,000 lb/30,845 kg.

GENERAL DYNAMICS F-16A FIGHTING FALCON

TYPE: *single-seat air combat and close support fighter*
WEIGHTS: *empty 17,780 lb/8,065 kg; maximum take-off 35,400 lb/16,057 kg*
DIMENSIONS: *span 31 ft 0 in/9.45 m; length 49 ft 4⅞ in/15.09 m; height 16 ft 8½ in/ 5.09 m; wing area 300 sq ft/27.87 m²*
POWERPLANT: *one 25,000-lb/11,340-kg afterburning thrust Pratt & Whitney F100-PW-200 turbofan*
PERFORMANCE: *speed 1,320+ mph/ 2,124+ km/h; ceiling 50,000+ ft/15,240+ m; range 1,150+ miles/1,850+ km*
ARMAMENT: *one 20-mm Vulcan multi-barrel cannon and 20,450 lb/9,276 kg of disposable stores (including six Sidewinder short-range air-to-air missiles, six Maverick air-to-surface missiles or two Harm anti-radar missiles plus a wide assortment of offensive and defensive electronic pods, rocket pods and both guided and unguided free-fall ordnance) on two wingtip missile rails, one underfuselage hardpoint and six underwing hardpoints*

LEFT The second YF-16 prototype on a test flight from the General Dynamics airfield at Fort Worth, Texas, in 1974.

BELOW LEFT The elegant and innovative lines of the first YF16 prototype; the most immediately evident features are the virtually variable-camber wings, the excellent field of vision for the pilot, and the large inlet for engine air under the fuselage.

RIGHT Heavyweight partner to the F-16, the McDonnell Douglas F-15 is designed for air superiority rather than air combat, and possesses quite exceptional performance.

BELOW The F-15 can carry an enormous payload in its secondary attack role, but is best used to secure air mastery with its 20-mm cannon and missile armament of AIM-7 Sparrow medium-range missiles (to be replaced by AIM-120 AMRAAMs) and AIM-9 Sidewinder short-range missiles.

GENERAL DYNAMICS F-16 FIGHTING FALCON IN FOREIGN SERVICE

The General Dynamics F-16 Fighting Falcon was schemed originally as a technology demonstrator for the US Air Force's lightweight fighter competition of 1974. However, its potential as a world-beating combat aircraft was such that it was ordered into production, making its service debut in 1978. Although designed as an air-combat fighter of exceptional agility, it has matured as a superlative all-round combat aircraft equally capable in the attack and close-support and the air-combat roles. Several improved models, both experimental and operational, have been produced and there seems every possibility that the Fighting Falcon will continue to appear in new guises for many years to come.

Given its capabilities, the F-16 has inevitably attracted considerable export interest from US allies and adherents, though the model specifically developed for this role – the F-16/79, with a General Electric J79 turbojet in place of the standard Pratt & Whitney F1100 turbofan – failed to attract significant interest. Even so the basic model has been exported to Belgium, Denmark, Egypt, Greece, Israel, the Netherlands, Norway, Pakistan, Singapore, South Korea, Thailand, Turkey and Venezuela.

Most exports involve standard aircraft, but Israel and Japan are developing upgraded models. Already an F-16 operator when its own IAI Lavi multi-role fighter was cancelled in 1987, Israel has decided to procure additional F-16s incorporating some of the Lavi's systems. Japan has opted for an improved variant with a bigger wing, fully modernized electronics and an element of composite construction. Perhaps most significantly of all, direct-force manoeuvring foreplanes under the inlet trunk, like those pioneered by the F-16/AFTI, will enhance the Japanese Agile Falcon's air-combat and attack manoeuvrability. The bigger wing of the Agile Falcon was originally proposed by General Dynamics to restore the agility lost by the heavier F-16C.

● and the wing is faired into the forward fuselage by highly-swept forebody strakes – often called lerxes, an abbreviation of leading-edge root extensions – to improve the formation of upper-surface vortices and improve handling at high angles of attack.

For practical purposes, therefore, the F-16's wing is of the variable-camber type, which means that it can be shaped in section to suit the five main flight regimes:
● takeoff and landing, with the leading-edge flaps angled up at 2° and the flaperons angled down at 20°;
● initial climb, with the leading-edge flaps angled down at 15° and the flaperons angled down at 20°;
● high-speed flight, with the leading-edge flaps angled up at 2° and the flaperons angled up at 2°;
● manoeuvring, with the leading-edge flaps angled down at 25° and the flaperons level;
● and approach, with the leading-edge flaps down at 15°, flaperons down at 20°.

LEFT The F-16 was designed after the lessons of the South-East Asia War, and was thus not optimized for maximum performance but rather for maximum 'flyability' in the air-combat arena – great agility, high rates of acceleration in all dimensions, and a cockpit offering the pilot unrivalled fields of vision.

The F-16's importance is attested by the number of production lines devoted to its manufacture, firstly in the USA (here) and then in a number of other countries.

The net effect of a fixed central section plus automatically scheduled movable leading- and trailing-edge sections is a wing that can be tailored to the particular conditions at only a modest cost in complexity and weight.

FLY-BY-WIRE CONTROL

The whole system of fixed and moving controls is operated by the F-16's digital flight-control system, often called fly-by-wire. This system interprets inputs from the pilot's force-sensing sidestick controller and rudder pedals to determine his intention. It then produces the optimum combination of control-surface deflections to implement the desired manoeuvre, acting in concert with an automatic aileron/rudder interconnect and yaw-rate limiter. The pilot has a seat inclined at 30° and a single-piece canopy affording excellent fields of view, his semi-reclining position allowing him to withstand higher g forces than is possible in the more conventional upright seat. The HOTAS (hands on throttle and stick) arrangement of

FAR LEFT Propellant gases stream from the muzzle of an F-16's M61A1 Vulcan cannon, buried on the port wing root but still an essential feature for dogfighting air combat at very short range.

LEFT The F-16's warload can include free-fall nuclear weapons, guided glide bombs and air-to-air missiles.

DASSAULT-BREGUET MIRAGE 2000 AND ARMAMENT

A keynote of modern combat aircraft design is flexibility of armament. A typical product of modern practice is the Dassault-Breguet Mirage 2000, which can lift up to 14,209 lb/6,445 kg of disposable stores on its external hardpoints – one centreline unit under the fuselage, two tandem units under each wing root and two units under each wing providing a total of nine weapon-carriage stations.

For air-to-air missions the Mirage 2000 can carry three medium-range Matra Super 530 interception and two short-range Matra R550 Magic dogfighting missiles, plus two internal 30-mm DEFA 554 cannon each supplied with 125 rounds of ammunition. In the air-to-surface role the type is hampered aerodynamically by its large, gust-prone wing, but can carry a wide assortment of useful weapons. Typical ground-attack weapons are a pair of AS.30L laser-homing air-to-surface missiles complemented by a pair of CC421 pods, each carrying a single 30-mm DEFA 554 cannon plus ammunition. Point targets can also be engaged with the Matra laser-guided bomb, an 882-lb/400-kg weapon of which seven can be carried.

Other weapons, less accurate but offering good area-attack potential are:
- the Beluga or Brandt 882-lb/400-kg modular cluster bomb;
- the BAP 100 runway-cratering bomb;
- the BAT 120 anti-vehicle bomb;
- free-fall bombs in various sizes;
- and pods for 2.68-in/68-mm or 3.94-in/100-mm unguided rockets.

The hardpoints can also carry drop tanks or napalm, and at least one is generally used for a podded electronic warfare system.

the cockpit, in which the pilot controls the throttle with his left hand and the sidestick controller with his right hand, enables both arms to be supported, another factor that allows the fly-by-wire system to be fixed at limits of 9 g and 26° angle of attack for conventional operations: it is the pilot's ability to tolerate g forces rather than the structure and/or aerodynamics of the airframe that has now become the limiting factor in air combat.

The European combat aircraft that comes closest to the F-16's CCV concept is the Dassault-Breguet Mirage 2000, which also uses a fly-by-wire control system and wings of simple variable-camber type. In configuration the Mirage 2000 is entirely unlike the F-16, being a tailless delta. Nevertheless, this important French fighter has an altogether higher level of flight agility than its similarly configured predecessor, the Mirage III, since CCV technology allows it to avoid the massive trim drag problems that so eroded the Mirage III's performance in manoeuvring and low-level operations.

MISSION-ADAPTIVE WINGS

Of course, it is better to have a wing that can be tailored over its whole chord to the best shape for any particular set of flight conditions, and such technology is now being pioneered by the mission-adaptive wing (MAW). Designed to be flexed in profile like that of a bird, it has been under development by Boeing since 1979 on the basis of the General Dynamics F-111 strike aircraft, which has the additional bonus of a variable-geometry wing planform.

The trouble with existing variable-camber wings is that the combination of one fixed and two or more movable portions results in dis-

ABOVE The first Dassault-Breguet Mirage 2000B shows off its in-flight refuelling capability.

LEFT One of the keys to the Mirage 2000's capabilities is the combination of a fly-by-wire control system with an advanced wing designed with moveable surfaces on both the leading and trailing edges.

BELOW LEFT A Mirage 2000 shows off part of its external load, in the form of two 374-Imp gal/1700-litre drop tanks, two Magic air-to-air missiles and eight 551-lb/ 250-kg free-fall bombs.

GENERAL DYNAMICS FB-111A

TYPE: two-seat medium-range operational/strategic bomber with variable-geometry wings
WEIGHTS: maximum take-off 114,300 lb/51,846 kg
DIMENSIONS: span 70 ft 0 in/21.34 m spread and 33ft 11 in/10.34 m swept; length 73 ft 6 in/22.40 m; height 17 ft 1 3/8 in/5.22 m
POWERPLANT: two 20,350-lb/9,231-kg afterburning thrust Pratt & Whitney TF30-PW-7 turbofans
PERFORMANCE: speed 1,650 mph/2,655 km/h; ceiling 60,000 ft/18,290 m; range 2,925+ miles/4,707+ km
ARMAMENT: up to 37,500 lb/17,010 kg of disposable stores carried in a small internal weapons bay and on six swivelling underwing hardpoints; a typical load is 42 750-lb/340-kg free-fall bombs, six free-fall thermonuclear bombs or six SRAM nuclear defence-suppression missiles, plus a wide assortment of electronic countermeasures

OPPOSITE, TOP The General Dynamics F-111's wing-sweep capability.

OPPOSITE, MIDDLE An FB-111A medium-range strategic bomber in cruise configuration with its bomb bay doors open to reveal two AGM-69A SRAM defence-suppression missiles.

LEFT The complexity of modern aerodynamics (both configuration and controls) is readily apparent in this illustration of an F-111D.

OPPOSITE, BOTTOM F-111s generally cruise with their wings unswept to maximize range, penetrating enemy airspace in the swept configuration for maximum speed and comfort of ride at low level.

BELOW End of day, and an F-111 taxies in with the canopies to its side-by-side seats open.

ABOVE From nose to tail the basically cylindrical fuselage of the Lockheed F-104G Starfighter accommodates electronics, the cockpit, fuel and the mighty afterburning turbojet.

ABOVE RIGHT Any view from below emphasizes the tiny and unswept wings of the F-104G and their Starfighter variants.

BELOW RIGHT Any photograph of a Starfighter showing the fuselage and wings explains why the type was dubbed the 'manned missile'.

continuities at the junctions. The MAW, on the other hand, provides smooth flexibility and the possibility of far more exact profile tailoring to suit operating conditions. Boeing's first move toward this concept was the leading-edge slats on its Model 747 airliner: these have flexible glassfibre skins, so they can be deployed in a smooth curve without discontinuity.

The MAW takes the process to its logical conclusion by enabling the whole wing to be flexed in profile hundreds of times per second. The structure is based on a rigid box attached to the wing-pivoting mechanism; to this are attached completely flexible leading- and trailing-edge sections with laminated glassfibre skins and a complex arrangement of internal actuators controlled by the aircraft's fly-by-wire system. The tactical advantages of such a wing are obvious, and even the prototype installation should allow range to be increased by up to 35% and provide a 25% improvement in sustained turn rate. In the longer term the MAW clearly offers both military and civil aircraft enormous advantages in performance and operating costs.

Until quite recent years, the emphasis in high-performance wing design had been on reducing thickness-chord ratio – that is the ratio of a wing's thickness to its breadth from leading to trailing edge – in order to reduce profile drag. Thin wings were common, especially on advanced military aircraft, though their use often meant that fuel and the main landing gear units had to be accommodated in

---

LOCKHEED F-104G STARFIGHTER

TYPE: single-seat multi-role and interdiction fighter
WEIGHTS: empty 14,900 lb/6,758 kg; maximum take-off 28,779 lb/13,054 kg
DIMENSIONS: span 21 ft 11 in/6.68 m; length 54 ft 9 in/16.69 m; height 13 ft 6 in/4.11 m; wing area 196.1 sq ft/18.22 m²
POWERPLANT: one 15,800-lb/7,167-kg afterburning thrust General Electric J79-GE-11A or MAN/Turbo-Union J79-MTU-J1K turbojet
PERFORMANCE: speed 1,450 mph/2,333 km/h; ceiling 58,000 ft/17,680 m; range 1,550 miles/ 2,495 km
ARMAMENT: one 20-mm Vulcan multi-barrel cannon and up to 4,310 lb/1,955 kg of disposable stores (including two Sidewinder short-range air-to-air missiles on the wingtip launcher rails and a wide assortment of conventional or – occasionally – nuclear free-fall and powered stores, both guided and unguided) on one underfuselage and four underwing hardpoints

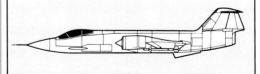

ABOVE The North American F-100 Super Sabre was a prodigious aerodynamic and structural achievement.

RIGHT Designed as an interceptor, the F-100 soon matured as a real workhorse of a fighter.

BELOW As it turns off onto a taxiway, this F-100 still streams its braking parachute.

the fuselage rather than in the wings. There was no problem with an aircraft like the Lockheed F-104 Starfighter, which has only a small and essentially unswept wing. However, such types as the North American F-100 Super Sabre and the contemporary Mikoyan MiG-19 were designed for greater agility and accordingly had bigger, highly swept wings which had low wave drag but involved the problem of aeroelastic distortion (that is, wing twist induced by reaction to aileron movement).

The designers of the F-100 skated round the problem by locating the ailerons inboard and omitting flaps, while the MiG-19 is a marvel of aircraft structure with ailerons and flaps. Other designers opted to fill in the area behind the swept trailing edge to create a delta wing of greater strength and internal volume, and one whose increased chord allowed greater depth for the same thickness/chord ratio. The classic example of this philosophy is the Mirage III, which has excellent straight-line performance at high altitude, but in sustained manoeuvres loses energy extremely rapidly as the large area of the wing is presented to the airflow.

LEFT Its large wing gives the Dassault Mirage III good performance at high altitude, but severely hampers its utility at low level.

BELOW LEFT The clean lines and massive delta wing of a Mirage IIIC interceptor; it carries two drop tanks and a single R530 air-to-air missile.

When flying at low levels, aircraft with delta or very slender wings exhibit a high response to gusts – local up- and down-currents that alter the wing's apparent angle of incidence. The resulting bumpiness in flight seriously curtails the life of the airframe and, more immediately, rapidly erodes its crew's ability to concentrate. Gust response is experienced by the crew as jerking vertical movements. The Mirage III suffers about 70 such 0.5-g bumps per minute, seven times the tolerable level, which can be achieved only by low-aspect-ratio aircraft such as the F-111 and Panavia Tornado. Yet the ability to fly for prolonged periods at low and very low levels is essential for modern aircraft faced with the task of penetrating hostile airspace.

The MAW not only has good gust-response characteristics, it is also ideally suited to the latest development in aerofoil section, namely the supercritical wing section. The supercritical wing is characterized by a fairly bluff leading edge, a bulged underside, a flat top and a down-curved trailing edge. The aerodynamics of such a section are complex, but they delay

LEFT The fairing under the rear fuselage of the Mirage III was designed to accept an SEPR liquid-propellant booster rocket, but the importance of high-altitude operations declined so rapidly in the late 1950s that this provision has seldom been employed.

RIGHT Though unsuccessful at the time, the Northrop YB-49 is now seen as a truly prophetic design – it is widely accepted that the new Northrop B-2 strategic bomber will have many conceptual similarities to the B-49 in layout.

maximum acceleration and shock formation, allowing the designer to create a wing with less sweep and a thicker section, which in turn provides extra volume and reduces structure weight. Supercritical wing sections are now used almost universally on military and civil aircraft designed for high subsonic and supersonic performance.

VARIABLE GEOMETRY

The MAW is also well suited to a variable-geometry layout. Variable geometry was pioneered in Germany during World War II, and further developed on an experimental basis in the USA during the late 1940s and early 1950s, but reached a practical level only with the F-111, which first flew in 1964.

It has been employed subsequently on several long-range combat aircraft, most notably the Tornado interdiction aircraft, the Rockwell B-1 penetration bomber, the Grumman F-14 Tomcat fleet-defence fighter and most new Soviet strike aircraft, among them the MiG-23/27 Flogger, Sukhoi Su-24 Fencer and Tupolev Blackjack.

NORTHROP YB-49

During development of the Northrop B-2 it was generally assumed that the US Air Force's new strategic stealth bomber would be a flying wing, and this would accord fully with the company's long association with this type of flying machine. The advantages of the flying wing over conventional aircraft stem from the enormous reductions in weight and drag resulting from the elimination of the fuselage and empennage (tail unit) and include reduced costs and much improved performance.

Jack Northrop had been an advocate of flying wings in the 1930s, and during World War II persevered with the ultimately unsuccessful XP-56 pusher fighter, which retained a vestigial fuselage. Greater success attended Northrop's larger flying wings, first of which was the XB-35 prototype strategic bomber. The XB-35 was exceptionally clean in line, being a true flying wing with tricycle landing gear; it had a range of 2,500 miles/4,023 km with a bombload of 20,000 lb/9,072 kg on the power of four 3,250-hp/2,424-kW Pratt & Whitney Wasp Major radials driving pusher propellers.

From the XB-35 Northrop developed the YB-49 jet bomber: powered by eight 4,000-lb/1,814-kg thrust Allison J35 turbojets, the YB-49 turned the scales at a maximum 216,600 lb/98,250 kg and reached a top speed of 520 mph/837 km/h, compared with the XB-35's 393 mph/632 km/h. Range was an excellent 2,800 miles/4,506 km with a 10,000-lb/4,536-kg bombload; an alternative bombload of 37,400 lb/16,965 kg could be carried over shorter ranges. However, there were problems with the design, and ultimately the production contract was cancelled.

(pictured opposite)

ABOVE The Rockwell B-1A was developed at a time when aviation technologies were evolving rapidly, and considering its cancellation with hindsight, it is possible to see that President Carter's much-vilified decision was in all probability the right one.

RIGHT The variable-geometry B-1A offered superb capabilities, but only at altitudes at which it would have been vulnerable to a whole host of sophisticated weapons.

ABOVE The value of the B-1A lies in the fact that it has paved the way for the B-1B operational bomber. This is still plagued with structural and avionics problems, but offers the real possibility of deep penetration into enemy airspace at high subsonic speed and very low levels: and by comparison with the B-1A the type has considerably lower radar signature to make its detection that much more difficult.

SUKHOI Su-7BMK FITTER-A

TYPE: single-seat close-support fighter
WEIGHTS: empty 19,004 lb/8,620 kg; maximum take-off 29,762 lb/13,500 kg
DIMENSIONS: span 29 ft 3½ in/8.93 m; length 57 ft 0 in/17.37 m including nose probe; height 15 ft 0 in/4.57 m; wing area 339.1 sq ft/31.50 m²
POWERPLANT: one 22,046-lb/10,000-kg afterburning thrust Lyul'ka AL-7F-1 turbojet
PERFORMANCE: speed 1,055 mph/1,700 km/h; ceiling 49,700 ft/15,150 m; range 430 miles/690 km
ARMAMENT: two 30-mm NR-30 cannon and up to 5,511 lb/2,500 kg of disposable stores (including a single tactical nuclear weapon, rocket pods, cannon pods, air-to-surface missiles and a wide assortment of free-fall conventional or cluster bombs) on two underfuselage and four underwing hardpoints

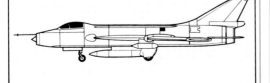

SUKHOI Su-17 FITTER-C

TYPE: single-seat close-support fighter with variable-geometry wings
WEIGHTS: empty 24,030 lb/10,900 kg; maximum take-off 39,020 lb/17,700 kg
DIMENSIONS: span 45 ft 11¼ in/14.0 m spread and 34 ft 9½ in/10.60 m swept; length 63 ft 0 in/19.20 m including nose probe; height 17 ft 6⅝ in/5.35 m; wing area 431.65 sq ft/40.10 m² spread and 400.4 sq ft/37.20 m² swept
POWERPLANT: one 24,691-lb/11,200-kg afterburning thrust Lyul'ka AL-21F-3 turbojet
PERFORMANCE: speed 1,432 mph/2,305 km/h; ceiling 59,055 ft/18,000 m; range 780 miles/1,255 km
ARMAMENT: two 30-mm NR-30 cannon and up to 8,818 lb/4,000 kg of disposable stores (including a single tactical nuclear weapon, rocket pods, cannon pods, air-to-surface missiles and a wide assortment of free-fall conventional or cluster bombs) on four underfuselage and four underwing hardpoints

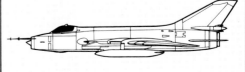

A typical variable-geometry, or swing-wing planform, such as that of the F-111, is based on a comparatively wide centre section accommodating the wing pivots and sweep actuators behind sharply swept wing gloves. Outboard of the pivots are the moving wing panels, which can be swept from a fully-forward 16° sweep angle to a maximum of 72.5°. In the forward positon the wings provide good low-speed lift and load-carrying capability, offering long range and the possibility of operations from short runways. When fully swept they provide low wave drag and a planform optimized for supersonic flight, as well as low gust response in high-speed low-level flight. The Western aircraft with the best gust response characteristics are the Tornado (bearable), the F-111 and B-1 (bearable for short periods) and the F-14 (on the verge of unacceptable); there is no reason to doubt that Soviet aircraft of this type enjoy the same benefits.

STOL CAPABILITY

Swing wings can also accommodate high-lift devices. Such devices are very important in increasing wing area and lift in low-speed flight, thereby providing short take-off and landing (STOL) capability.

This is an increasingly important feature of modern combat aircraft, though Western

ABOVE LEFT The Sukhoi Su-24 is the Soviet equivalent of the American F-111, like the US aircraft offering the possibility of long-range interdiction with its good avionics and variable-geometry layout.

ABOVE The MiG-23 is another variable-geometry Soviet aeroplane and a versatile multi-role fighter.

BELOW The MiG-23's ventral tail folds sideways on the ground to avoid contact with the runway.

ABOVE The use of a shoulder-set wing means that the MiG-23's main landing gear units are attached to the lower fuselage, with a somewhat complex arrangement to provide a wide enough track to avoid instability problems on the runway.

MIKOYAN MiG-23MF FLOGGER-B

TYPE: single-seat air-combat and multi-role fighter with variable-geometry wings
WEIGHTS: empty 24,250 lb/11,000 kg; maximum take-off 41,667 lb/18,900 kg
DIMENSIONS: span 46 ft 9 in/14.25 m spread and 26 ft 9½ in/8.17 m swept; length 59 ft 6½ in/18.15 m including the nose probe; height 14 ft 4 in/4.35 m; wing area 293.4 sq ft/27.26 m²
POWERPLANT: one 25,353-lb/11,500-kg afterburning thrust Tumanskii R-29 turbojet with variable-geometry inlets and variable nozzle
PERFORMANCE: speed 1,522 mph/2,450 km/h; ceiling 60,040 ft/18,300 m; range 1,180 miles/1,900 km
ARMAMENT: one 23-mm GSh-23L twin-barrel cannon and up to 6,614 lb/3,000 kg of disposable stores (including six air-to-air missiles and an extremely wide assortment of disposable stores such as air-to-surface missiles, rocket pods and free-fall conventional or nuclear bombs) carried on one underfuselage, two under-trunk and two underwing hardpoints

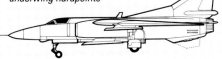

MIKOYAN MiG-27 FLOGGER-D

TYPE: single-seat close-support and ground-attack fighter with variable-geometry wings
WEIGHTS: empty 23,778 lb/10,790 kg; maximum take-off 44,313 lb/20,100 kg
DIMENSIONS: span 46 ft 9 in/14.25 m spread and 26ft 9¾ in/8.17 m swept; length 52 ft 5⅞ in/16.0 m; height 14 ft 9.2 in/4.50 m; wing area 293.33 sq ft/27.25 m² spread
POWERPLANT: one 25,353-lb/11,500-kg afterburning thrust Tumanskii R-29 turbojet with plain inlets and two-position nozzle
PERFORMANCE: speed 1,123 mph/1,807 km/h; ceiling 52,495 ft/16,000 m; range 480 miles/780 km
ARMAMENT: up to 8,818 lb/4,000 kg of disposable stores (typically a wide assortment of air-to-surface weapons of the unguided free-fall conventional or nuclear and guided powered types, plus rocket pods) carried on three underfuselage, two under-trunk and two underwing hardpoints

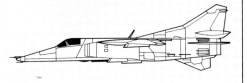

SAAB 37 VIGGEN OFF-RUNWAY DISPERSAL

Just about every air base in western Europe is a likely target for Soviet attack in the event of war: it is probable that within hours if not minutes of the outbreak of hostilities every substantial runway would be destroyed by a nuclear attack or severely cratered by conventional weapons. Western European combat aircraft, together with American and Canadian machines based on the eastern side of the Atlantic, would be unable to operate even if they had survived the attack in their hardened aircraft shelters. Cratered runways can be repaired, but they can be re-cratered just as easily, and in such circumstances the only Western aircraft still able to operate would be the British Aerospace Harrier, with its short take-off and vertical landing capability, and the Swedish Saab 37 Viggen.

The Viggen is the aircraft component of the Swedes' System 37 defence scheme, and was designed to operate without loss of capability from dispersed sites, comprising 545-yard/500-m lengths of straight road in any part of the country. System 37 also involves large numbers of vehicles to ensure that maintenance crews, along with fuel and ordnance, can be ferried to these dispersed sites as required in times of crisis. To ease the maintenance load at such sites the Viggen has been designed for maximum reliability and maintainability, with a light airframe stressed to 12 g and ground-level access to most major systems in the airframe.

The Viggen's canard configuration is admirably suited to steep, slow arrivals and departures at the dispersed sites, and its landing gear is built by Motala Verstad to meet extremely exacting requirements for no-flare landing at a sink rate of 16.4 ft/5 m per second. Each main unit features a long-stroke oleo, shortened during retraction to require less internal volume, and a tandem arrangement of two wheels pressurized to 215 lb/sq in 15.12 kg/cm^2 to help absorb the considerable landing shock. As soon as the twin nosewheels hit the ground, the compression of the nose unit oleo activates the switch that operates the engine thrust-reverser system. (The Viggen's powerplant is a Volvo Flygmotor RM8A afterburning turbofan, derived from the civil Pratt & Whitney JT8D for maximum reliability.) As soon as the switch is operated, the engine's exhaust is deflected forward through three annular slots in the rear fuselage, complementing the brakes to halt the Viggen safely and quickly.

countries have been slow to accept a fact that has been evident to the Soviets and Swedes for decades. Almost all Western military aircraft depend on expensive air bases with extensive hangarage and vast concrete runways which are conspicuously vulnerable to conventional weapons, leaving aside the effect of a nuclear attack.

It is estimated that all NATO's airfields have been targeted by at least two Soviet nuclear-armed missiles, raising the very real possibility that once a nuclear exchange had been initiated all NATO air bases would either cease to exist or, at best, be severely damaged. Conventional aircraft cannot operate from short stretches of damaged runway; so there is every likelihood that the vast majority of NATO's tactical aircraft would be rendered impotent, even if their hardened aircraft shelters protected them from the initial strike.

The Swedes have opted for STOL aircraft, in the shape of the Saab 35 Draken and Saab 37 Viggen, which can operate from dispersed sites such as straight stretches of country road. The Soviet answer is a generation of tactical aircraft whose high-lift devices and landing gear are suitable for rough field operations. Some

SAAB AJ37 VIGGEN

TYPE: single-seat all-weather attack aircraft
WEIGHTS: empty 26,015 lb/11,800 kg; maximum take-off 45,194 lb/20,500 kg
DIMENSIONS: span 34 ft 9¼ in/10.60 m for main delta wing and 17 ft 10½ in/5.45 m for canard foreplane; length 53 ft 5¾ in/16.30 m; height 19 ft 0¼ in/5.80 m; wing area 495.1 sq ft/46 m^2 for main delta wing and 66.74 sq ft/6.20 m^2 for canard foreplane
POWERPLANT: one 26,015-lb/11,800-kg afterburning thrust Volvo Flygmotor RM8A turbofan (licence-built Pratt & Whitney JT8D-22 with locally designed afterburner and thrust reverser)
PERFORMANCE: speed 1,320 mph/2,125 km/h; ceiling 49,870 ft/15,200 m; range 1,243+ miles/2,000+ km
ARMAMENT: up to 13,228 lb/6,000 kg of disposable stores (including air-to-air missiles, air-to-surface missiles, cannon pods, rocket pods and various types of free-fall bomb) on three underfuselage and four underwing hardpoints

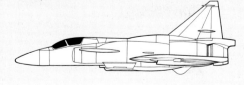

NATO countries are finally coming to grips with the problem too, and the Americans are currently investigating a derivative of the F-15 Eagle whose two-dimensional thrust-vectoring system provides the ability to operate from 500-yard/455-m lengths of undamaged runway.

HIGH-LIFT TORNADO

The NATO type best able to deal with damaged runways is the swing-wing Tornado, whose primary control surfaces are slab tailplane halves that operate in unison for pitch control and differentially for roll control and a powerful rudder for yaw control. The wings are left free

ABOVE For its time the Saab 35 Draken was a highly innovative design, the double-delta wing offering great lift and large fuel volume without a massive drag penalty. This is the J 35E reconnaissance version.

RIGHT The Saab 37 Viggen is another innovative design from Sweden, this time using a canard configuration and thrust-reversing turbofan for exceptional short-field performance.

LEFT The SF 37 is the overland reconnaissance version of the Viggen, external reconnaissance equipment being used to provide the maximum flexibility of sensor carriage.

RIGHT The Panavia Tornado GR Mk 1 is in its element at high speed and very low level.

BELOW The real skill in designing modern combat aircraft lies not just in producing a machine capable of meeting its specification, but of meeting the specification in as compact an airframe as possible. The Tornado is a superb example of such skill, yet has considerable growth potential.

FAR RIGHT Below the nose of this Tornado GR Mk 1 is the fairing over the laser ranger and marked-target seeker.

BELOW RIGHT This Tornado GR Mk 1 carries two JP233 airfield attack dispensers, each carrying runway-cratering submunitions and area-denial minelets to hamper the repair crews.

PANAVIA TORNADO IDS

TYPE: two-seat all-weather interdiction and multi-role fighter with variable-geometry wings
WEIGHTS: empty 31,063 lb/14,090 kg; maximum take-off 60,000 lb/27,215 kg
DIMENSIONS: span 45 ft 7½ in/13.91 m spread and 28 ft 2½ in/8.60 m swept; length 54 ft 10¼ in/16.72 m; height 28 ft 2½ in/5.95 m; wing area about 269 sq ft/25.0 m^2
POWERPLANT: two 16,800-lb/7,620-kg afterburning thrust Turbo-Union RB 199-34R Mk 103 turbofans
PERFORMANCE: speed 1,453+ mph/ 2,337+ km/h; ceiling 49,210+ ft/15,000 m; range 1,727+ miles/2,780+ km
ARMAMENT: two 27-mm Mauser BK27 cannon and up to 19,840 lb/9,000 kg of disposable stores (including Sidewinder short-range air-to-air missiles, Kormoran or Sea Eagle anti-ship missiles, Maverick air-to-surface missiles, guided glide bombs, unguided free-fall conventional or cluster bombs, dispenser weapons, rocket pods and a wide assortment of electronic countermeasures pods) on three underfuselage and four swivelling underwing hardpoints

for the high-lift devices that are claimed to give the Tornado the best lifting qualities of all supersonic aircraft with its wings in the 25° minimum-sweep position.

The leading edge of each Tornado wing is occupied over its full span by three-section leading-edge slats. The trailing edge has full-span, four-section, double-slotted flaps complemented by two-section lift-dumpers which also operate as spoilers to supplement the tailerons in roll control. And the high-lift system is completed by a Krüger flap on each 60° swept wing glove. Widely used on civil aircraft, the Krüger flap is a leading-edge device forming part of the undersurface of the leading edge, being hinged to swing down and forward to create a bluff leading edge suitable for low airspeeds on a high-speed wing. The whole system is enhanced by a powerful airbrake on each side of the vertical tail, and by thrust reversers on the two afterburning turbofan engines.

Such high-lift devices are standard on civil and military transport aircraft, which may have triple-slotted flaps complemented by drooping ailerons. Lift is enhanced by the increasing lifting area of the extended flaps, which also control the airflow for optimum lift at low airspeeds, while drag is generated by the additional area and deflected position of the flaps in order to lower the aircraft safely onto the runway at the minimum possible airspeed. Safety is thereby increased, while runway requirements are minimized.

Some airliners are also provided with winglets, devices which are set to become more common on long-range military aircraft such as strategic transports and maritime patrol aircraft. The term is a misnomer, since winglets are actually upturned – or occasionally downturned – wingtip extensions designed to improve the cruise efficiency of the wing. They do so by reducing the tip vortex, consequently curtailing induced drag and recovering the energy otherwise lost to this cause, and by enhancing the circulation of air over the outer wing panels to generate additional lift.

ABOVE The classic configuration of high-lift devices on transport aircraft is revealed on this Boeing 727-200 airliner, which sports Krueger leading-edge flaps (inboard) and leading-edge slats (outboard) as well as triple-slotted trailing-edge flaps.

RIGHT A test model of the McDonnell Douglas C-17 long-range transport, currently under development for the US Air Force, reveals the configuration of the modern tactical transport, with a T-tail, massive fuselage accommodating the main landing gear units in blister fairings, and an uncluttered wing with winglets.

WINGLETS

Developed largely on the basis of research by Richard T. Whitcomb, one of the most prolific of modern aerodynamicists, the winglet is a highly effective way of boosting the cruise efficiency of a wing. The winglet comprises an extension of the main wing at the tip; turned either up or down, it serves the double purpose of reducing the wingtip vortex, thereby recovering the energy that would otherwise be lost in the vortex, and improving the circulation – and therefore the lift – of the outer wing panel. The wingtip vortex is a high-energy rotational movement of air which streams back, out and generally down from the tip, producing considerable drag: generally fitted over the rear half of the tip chord and angled outward slightly, the winglet serves both to reduce and to control this vortex.

The winglet also helps prevent spanwise separation of the airflow over the upper surface of the wing – a similar purpose is served by the prominent fences on the wings of many Soviet aircraft – and reduces the leakage of higher-pressure air from the under surface of the wing round the tip to the lower-pressure air flowing over the upper surface. As this pressure differential is the feature that creates lift, its maximization by the winglet, allied to other devices, allows overall area and structural weight to be reduced.

VTOL AND STOVL

The problem of runway vulnerability can be avoided entirely by vectoring – that is, pointing – engine thrust to permit vertical take-off and landing (VTOL) or, in more practical operational terms, short take-off and vertical landing (STOVL). The classic example of this type of aircraft is the British Aerospace Harrier, which is now in service in its much revised Anglo-American form as the McDonnell Douglas/British Aerospace Harrier II.

The rationale for such an aeroplane is unimpeachable: if the total thrust of the engine exceeds the take-off weight of the aeroplane, and if this thrust can be vectored downwards to provide direct lift, then the aeroplane can rise straight up from the ground without using a runway at all. VTOL aircraft can be based anywhere in times of crisis, virtually eliminating their vulnerability to pre-emptive destruction and adding enormously to their tactical versatility so long as fuel, ordnance and other supplies can be provided at their operating sites.

Inevitably, there is a price to be paid for the tactical advantages offered by VTOL. The most obvious drawback is that for vertical take-off the thrust/weight ratio must exceed unity, and this generally means that the Harrier can only lift off vertically with a reduced weapon load or a reduced fuel load. In theory this disadvantage can be overcome by the use of in-flight refuelling: the Harrier could take off with maximum weapons but only minimum fuel, and after translation from engine- to wing-borne flight could rendezvous with a tanker aircraft to fill its tanks. Such a scheme would tie the dispersed-site Harrier to runway-based tankers, however. Instead, the answer lies in a compromise between thrust- and wing-borne flight to allow take-off with useful weapon and fuel loads.

Operational practice involves a very short take-off run: with the engine nozzles at 0°, pointing straight aft for maximum thrust, the pilot accelerates for a short distance then pulls the thrust lever back to the 90° position. All the thrust is instantly converted into a vertical component that jumps the aeroplane into the air, where the pilot can ease the lever back towards the 0° position and so translate rapidly into wing-borne flight. Landing vertically presents no problem, for the weapon load will

ABOVE The basic concept of the British Aerospace Harrier has been taken one step further in the McDonnell Douglas AV-8B Harrier II, an advanced development most notable for its superior cockpit, larger single-piece wing of composite construction and supercritical section, and considerably enhanced lift-improvement devices. Two side strakes serve as the mountings for the 25-mm cannon and its ammunition supply.

BRITISH AEROSPACE HARRIER GR Mk 3

TYPE: single-seat close-support and reconnaissance aircraft with STOVL (short take-off and vertical landing) capability
WEIGHTS: empty 13,535 lb/6,139 kg; maximum take-off 25,200+ lb/11,431+ kg
DIMENSIONS: span 25 ft 3 in/7.70 m or 29 ft 8 in/9.04 m with low-drag bolt-on ferry tips; length 46 ft 10 in/14.27 m; height 11 ft 4 in/3.45 m; wing area 201.1 sq ft/18.68 m^2 or 216.0 sq ft/20.07 m^2 with bolt-on ferry tips
POWERPLANT: one 21,500-lb/9,752-kg thrust Rolls-Royce Pegasus II Mk 103 non-afterburning turbofan with four vectoring nozzles, the forward pair for cold gas from the fan and the aft pair for hot gas from the core
PERFORMANCE: speed 737+ mph/1,186 km/h; ceiling 50,000+ ft/15,240+ m; range 828 miles/1,316 km
ARMAMENT: two 30-mm Aden cannon and up to 8,000 lb/3,269 kg of disposable stores (including Sidewinder short-range air-to-air missiles, Martel air-to-surface missiles, rocket pods, Paveway laser-guided glide bombs and a wide assortment of free-fall conventional or cluster bombs) on one underfuselage and four underwing hardpoints

BRITISH AEROSPACE SEA HARRIER FRS Mk 1

TYPE: single-seat carrier-borne fighter, reconnaissance and strike aircraft with STOVL (short take-off and vertical landing) capability
WEIGHTS: empty 13,100 lb/5,942 kg; maximum take-off 26,190 lb/11,880 kg
DIMENSIONS: span 25 ft 3 in/7.70 m; length 47 ft 7 in/14.50 m; height 12 ft 2 in/3.71 m; wing area 201.1 sq ft/18.68 m^2
POWERPLANT: one 21,500-lb/9,752-kg thrust Rolls-Royce Pegasus II Mk 104 non-afterburning turbofan with four vectoring nozzles, the forward pair for cold gas from the fan and the aft pair for hot gas from the core
PERFORMANCE: speed 735 mph/1,183 km/h; ceiling 50,000+ ft/15,240+ m; range 920 miles/1,480 km
ARMAMENT: two 30-mm Aden cannon and up to 8,000 lb/3,629 kg of disposable stores (including Sidewinder short-range air-to-air missiles, Martel air-to-surface missiles, Harpoon or Sea Eagle anti-ship missiles, rocket pods, Paveway laser-guided glide bombs, the WE-177 tactical nuclear bomb and a wide assortment of free-fall conventional or cluster bombs) on one underfuselage and four underwing hardpoints

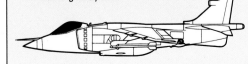

ABOVE The AV–8B in classic operating locale: any small space that can be cleared close to the front line.

RIGHT The British version of the AV–8B is the Harrier GR Mk 5, which has a number of detail modifications compared with the American model and which, like the AV-8B, can ultimately be fitted with radar for an all-weather attack capability.

By removing the need for a fixed operating base with its vulnerable runways, the designers of the STOVL Harrier created a new air weapon. The story of Harrier operations is encapsulated in this sequence of photographs:

LEFT The Harrier GR Mk 3 operates from concealed sites such as woodland clearings close to a road or track that can be used by the squadron's fuel, ordnance and other logistical vehicles.

BELOW LEFT The Harrier can operate from its clearing in the vertical take-off mode, but is able to uplift a heavier warload after a short take-off from a length of road.

BELOW RIGHT The Harrier can carry a substantial load, but has the performance and agility to cope with dedicated fighters mustered by the enemy.

RIGHT With its mission completed the Harrier lands vertically and immediately taxies off the strip into the concealment of the woods.

ABOVE The Harrier's simple airfield operations mean that the type can do more than adequately with small plain flaps, leaving the wing uncluttered by the normal plethora of high-lift devices, thus designed and built for flight.

LEFT Viewed from below, the Harrier reveals the massive fuselage required for the turbofan, its large inlets and the four-poster deflection system for the cold and hot gases via the front and rear nozzles respectively.

RIGHT An overhead shot of the AV-8B construction line highlights the one-piece wing, based on a composite-structure torque box accommodating a large volume of fuel.

have been expended and the fuel load greatly reduced, restoring the thrust/weight ratio to greater than unity.

Various expedients have been adopted in the designs of the Harrier, Sea Harrier and Harrier II to increase vertical lift. As much as possible of the gas vectored down from the engine is trapped, and control at nil or very low airspeeds is provided by reaction nozzles using cool air tapped from the engine compressor. In general, though, the use of a thrust-vectoring powerplant allows the flying surfaces to be much simpler than those of a comparable fixed-thrust type. Thrust-vectoring also makes for enormous agility in the air, and the Harrier variants have proved in many air-combat manoeuvres that they can use the technique known as VIFFing (vectoring in forward flight) to master even dedicated air-combat fighters like the F-16.

In combat, VIFFing is used for direct lift control comparable to that of experimental machines such as the AFTI/F-16. If the pilot of a conventional fighter finds an opponent on his tail, he may try to shake it off with a series of tight turns, but air-combat fighters generally have the agility to follow and even to close in such manoeuvres. In a comparable situation,

ISRAEL AIRCRAFT INDUSTRIES Kfir-C2

TYPE: single-seat interceptor and ground attack aircraft
WEIGHTS: empty 16,060 lb/7,285 kg; maximum take-off 35,714 lb/16,200 kg
DIMENSIONS: span 26 ft 11½ in/8.22 m for the main delta wing and 12 ft 3 in/3.73 m for canard foreplane; length 51 ft 4½ in/15.65 m including nose probe; height 14 ft 11¼ in/4.55 m; wing area 374.6 sq ft/34.80 m^2 for main delta wing and 17.87 sq ft/1.66 m^2 for canard foreplane
POWERPLANT: one 17,900-lb/8,119-kg afterburning thrust General Electric J79-GE-J1E turbojet
PERFORMANCE: speed 1,516+ mph/2,440+ km/h; ceiling 58,000+ ft/17,680 m; range 428 miles/690 km
ARMAMENT: two 30-mm DEFA 552/553 cannon and up to 12,731 lb/5,775 kg of disposable stores (including Shafrir 2 or Python 3 short-range air-to-air missiles, Luz-1 or Maverick air-to-surface missiles, Shrike anti-radiation missiles, rocket pods, guided glide bombs and a wide assortment of free-fall conventional or cluster bombs) on five underfuselage and four underwing hardpoints

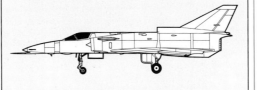

ABOVE Comparison with earlier illustrations of the F-16 confirms the radically different appearance of the AFTI/F-16, the result of the addition of the canted control surfaces under the inlet. Less obvious is the larger dorsal strake accommodating this research aeroplane's mass of additional avionics.

ABOVE RIGHT By adding canard foreplanes to its Kfir (itself a development of the Mirage III/5 series), IAI has produced the much enhanced Kfir-C2 with greater combat agility and far superior field performance.

MIDDLE RIGHT The Kfir-TC2 is the two-seat version of the Kfir-C2, the larger nose and other volumes being filled with additional electronics for the type's primary role of electronic warfare under the guidance of the officer in the rear seat.

BELOW RIGHT France's refusal to sell Israel the clear-weather Mirage 5 prompted full-scale development of the Kfir at a rapid pace; but the baseline Mirage 5 then went on to secure substantial orders from many nations. This example is in the markings of Venezuela.

the Harrier pilot can roll into the turn and select thrust 90° down: all the engine's thrust is then directed straight into the direction of the turn, reducing its radius to a degree unmatchable by a conventional fighter. And as the thrust-vectoring system allows a maximum 110° deflection (20° forward of straight down), the system can also be used to brake the Harrier more rapidly than a conventional fighter's airbrakes can manage.

The combination of a compact and inherently agile airframe with a thrust-vectoring system gives the STOVL aircraft an unprecedented level of agility. It is truly remarkable that the combination has not yet been adopted for dedicated air-combat fighters, which do not need to carry heavy offensive loads and could therefore afford to take off vertically.

CANARD FOREPLANES

The canard configuration, involving a tail-mounted delta wing and nose-mounted delta foreplanes, as used in the vast B-70 strategic bomber, is becoming standard for smaller fixed-wing combat aircraft. There are two good aerodynamic reasons for its application to such types as the Swedish Saab 37 Viggen and the Israel Aircraft Industries Kfir. Firstly, the nose of a normal tailless delta aircraft can be raised only by the application of up-elevon, which produces a down-load at the tail that must be subtracted from overall wing lift in any lift equation. This translates into a longer take-off run – since the landing gear is being

pushed down onto the runway, the aeroplane has to accelerate to a higher speed to compensate for what is effectively a smaller wing – and an increase in the radius of any turn, since the tail is being pushed away from the origin of the turn radius. However, with a canard configuration the take-off run is shortened, since the elevons on both the tail and nose deltas are deflected downwards, creating up-loads that must be added to the overall wing/canard lift; and the turning radius is reduced as the foreplane pulls the nose in towards the origin of the turn radius.

Secondly, the canards create powerful vortices that stream back and out over the upper surfaces of the wings, re-energizing the sluggish boundary-layer vortices of those surfaces and delaying the breakaway of the airflow over their all-important outer panels. This is particularly important at high angles of attack, and a similar though less effective result can be secured in aircraft of a more conventional configuration by the use of leading-edge vortex generators (small bladelike projections, otherwise called turbulators), dogtoothed leading edges, which also reduce the thickness/chord ratio of the outer panels for useful advantages at transonic speeds, and LERXes or fuselage chines.

The canard has now become extremely fashionable in combat aircraft circles. The configuration has been adopted for the Viggen's successor, the JAS 39 Gripen; it will be used for Europe's two next-generation fighters, the Dassault-Breguet Rafale and the multi-national European Fighter Aircraft; and it may well be featured by the US Air Force's Advanced Tactical Fighter, as well as the Soviet aircraft.

TRANSONIC TAILS

A prominent characteristic of modern combat aircraft is their large horizontal and vertical tail surfaces. On supersonic aircraft the former are generally of the fully-powered slab type, in which the whole surface moves to ensure adequate pitch control during the severe trim changes that occur at transonic speeds. Such control problems were only appreciated in the early 1950s as jet aircraft began to approach the speed of sound barrier; they were accentuated by the accompanying reduction in wing area, especially of swept-wing aircraft. Many aircraft were lost to Dutch rolling, a lateral oscillation of the aeroplane with increasingly uncontrol-

lable yaw and roll components. Larger tail surfaces did much to cope with the problem, and are particularly evident on Soviet aircraft: the MiG-25 has massive outward-canted vertical tail surfaces, while the MiG-23 is fitted with a ventral fin so large that it has to be folded sideways for take-off and landing.

DIRECT CONTROL MODES

Controllability is a key feature of modern aerodynamics, and lies at the heart of the CCV technology discussed above. But the standard controls of current aircraft can only rotate the aircraft around the relevant axis – longitudinal for roll, transverse for pitch and vertical for yaw – which means that it is some time before changes in one or more planes modify the aircraft's trajectory. This can waste a lot of time.

For example, if a pilot is about to attack a target with gunfire and finds that the crosswind has pushed his aircraft too far to one side, he has to roll in the right direction for what he estimates to be the correct distance, and then roll in the opposite direction to get back onto

OPPOSITE, BELOW Typical of the shape that will become standard in the 1990s is the Dassault-Breguet Rafale. This epitomizes the new generation of air-combat fighter with fly-by-wire controls, a close-coupled canard configuration, extensive use of composites, and advanced electronics of both offensive and defensive varieties.

OPPOSITE, ABOVE Another keynote of the Rafale's design is the elegance of its aerodynamic lines, which boost performance while helping to reduce radar cross-section.

ABOVE The USSR is also heavily committed to the new generation of blended aerodynamics, with aircraft such as this Mikoyan-Gurevich MiG-29, an air superiority and multi-role fighter possessing marked affinities with the McDonnell Douglas F-15 in design, but rather more like the same company's F/A-18 Hornet dual fighter and attack aircraft in size.

BELOW LEFT A view from below of the MiG-29 emphasizes the type's aerodynamic/structural blend, based on a wide fuselage/centre section with leading-edge root extensions, modestly swept wings, widely spaced underset engines and a powerful, all-moving tail.

the right trajectory. It would be much quicker if the pilot had additional control surfaces able to exert direct force normal to the trajectory: in the example quoted he could use his controls to exert a lateral force, shifting his aeroplane sideways without altering his heading until the target was right in his sight. Conversely, he could alter the lateral displacement of the nose without modifying his basic flight trajectory in order to keep the target in his sight while maintaining a course offset slightly to one side of it.

Similar techniques were pioneered in the form of direct lift control by airliners such as the Lockheed L-1011-500 TriStar, which use their spoilers to modify their glidepath, and have long been a feature of glider landings, where the lift is dumped to achieve a high rate of descent with the fuselage level. But the concept has been taken a stage further by the advanced fighter technology integration version of the Fighting Falcon.

The AFTI/F-16, has additional control surfaces in the form of two outward-canted oblique foreplanes located under the inlet duct and two outward-canted fixed ventral fins. The control surfaces are used in conjunction with a highly advanced digital flight-control system, and the aircraft has demonstrated unprecedented flight agility. In the pitch plane, for example, it can elevate its nose without gaining height and gain height without lifting its nose. Comparable modifications are attainable in the other two planes, and the most significant limitation so far encountered is the inability of the pilot to sustain accelerations of 2 g or more, normal to the line of flight, while remaining fixed in his seat and looking forward through the head-up display. Nevertheless, the AFTI/F-16 clearly points the way ahead for the aerodynamic control of high-performance aircraft.

CONFORMAL WEAPON CARRIAGE

Another radical development is the cranked-arrow F-16XL. The F-16's normal wing and tailplane have been replaced by a large cranked-arrow delta wing, which is essentially an ogival delta wing with extended tips: the wings curve out from a point under the cockpit at a sweep angle of about 50°, then angle back at 70° to a point as far outboard as the two structural beam fairings that project aft from the trailing edge, before continuing to the tip at the reduced sweep of 50°.

GENERAL DYNAMICS F-16/AFTI COCKPIT

Modified as part of the American Advanced Fighter Technology Integration (AFTI) programme, the General Dynamics F-16/AFTI flying testbed is immediately recognizable by the obliquely-canted canards under its inlet trunk and an enlarged dorsal spine housing additional electronics. The configuration had been pioneered with the YF-16/CCV (control-configured vehicle) adaptation of a pre-production Fighting Falcon, but the F-16/AFTI's triplex digital fly-by-wire flight-control system made possible fully decoupled (non-rotative) flight manoeuvres.

Flight trials confirmed that the system opened new ways to fly. In the vertical plane, for example, a conventional aircraft has to raise its nose to climb, whereas the F-16/AFTI can gain or lose height without altering its attitude, which could be useful during weapon delivery. Conversely, the F-16/AFTI can alter its angle of attack – the angle at which its wings meet the airflow – without rising or falling, a useful facility in air-to-air gunnery.

The cockpit of the F-16/AFTI is typical of the basic F-16, but has been modified to provide CCV monitoring. The most obvious modification is the special wide-angle head-up display developed in the UK by GEC Avionics. The standard F-16 HUD has a comparatively small field of view – 13.5° in azimuth and 9° in elevation – and was clearly unsuitable for the decoupled flight capability of the F-16/AFTI. The company therefore developed a more advanced HUD with 20° × 15° field of vision. The latter was used as the basis for the 30° × 20° HUD installed on F-16Cs and Ds, the holographic unit originally planned having proved impossible to produce.

LEFT The Lockheed TriStar is typical of modern airliner design so far as the powerplant is concerned; the two wing-mounted engines are suspended on projecting pylons where their mass helps to alleviate wing twist and bend. Such a location also allows easy maintenance of the engine and comparatively straightforward substitution of another engine (of either the same or a newer and probably larger type), and leaves the wing uncluttered to perform its primary aerodynamic functions.

BELOW Airliners such as the TriStar are models of careful design, commercially successful types blending safety and the last word in operating economics.

ABOVE Compared with the initial Tornado IDS variant, the Tornado ADV has a longer fuselage, and this pays handsome dividends in additional fuel capacity, a better fineness ratio for enhanced transonic acceleration, and the possibility of locating four Sky Flash air-to-air missiles in semi-recessed positions in the lower fuselage. This example also sports a pair of Sidewinder short-range missiles on the inner sides of the pylons for the drop tanks.

RIGHT The General Dynamics F-16XL is clearly a close relative of the standard F-16 Fighting Falcon fighter, but the use of a 'cranked-arrow' delta wing and a 56-in/1.42-m longer fuselage boosts internal fuel capacity by 80%, allowing the type to carry twice the payload over the same combat radius as the F-16, or the same payload over double the combat radius.

The F-16XL's basic flight performance remains roughly comparable with that of the standard F-16; where it really comes into its own is in weapon carriage, with a maximum of 29 hardpoints available on 17 stores stations. The stores stations are semi-conformal types to minimize drag, which is 58% less than that of the F-16, enabling the F-16XL to carry twice the payload 45% further than the basic F-16. The conformal carriage of weapons is already becoming important. External carriage of disposable weapons became common in World War II. It allowed obsolescent fighters to be fitted with bombs and rockets for ground-attack work, and current types to carry drop tanks as a means of extending range, and weapons have been carried externally ever since. It was realized all along that external stores produced a high level of drag, to the detriment of performance, but operational flexibility was deemed more important than the reduction of drag by the elimination of exterior pylons and their fittings. Some dedicated single-role aircraft, such as the B-52 heavy bomber and the Convair F-102 and F-106 interceptors, had internal weapons bays, but in general the accommodation of weapons inside the fuselage was deemed unnecessary.

LEFT Considerable attention is devoted to the maximum use of internal volume, the Common Strategic Rotary Launcher being a good example; it allows the carriage of eight weapons such as these AGM-86B air-launched cruise missiles, which remain trimly folded until they have been launched.

MIKOYAN MiG-21 FISHBED-K

TYPE: *single-seat dual-role fighter and light attack aircraft*
WEIGHTS: *empty 12,302 lb/5,580 kg; maximum take-off 20,723 lb/9,400 kg*
DIMENSIONS: *span 23 ft 5½ in/7.15 m; length 51 ft 8½ in/15.76 m including nose probe; height 13 ft 5½ in/4.10 m; wing area 247.6 sq ft/23.0 m²*
POWERPLANT: *one 14,550-lb/6,600-kg afterburning thrust Tumanskii R-13-300 turbojet*
PERFORMANCE: *speed 1,385 mph/2,230 km/h; ceiling 50,030 ft/15,250 m; range 460 miles/740 km*
ARMAMENT: *one 23-mm GSh-23L twin barrel cannon and up to 3,307 lb/1,500 kg of disposable stores (including four AA-2 Atoll or AA-8 Aphid short-range air-to-air missiles, cannon pods, rocket pods and free-fall bombs) on four underwing hardpoints*

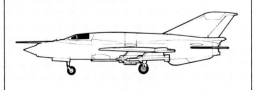

HUGHES AIM-54 PHOENIX

Undoubtedly the most capable air-to-air missile in the world, the Hughes AIM-54 Phoenix is carried only by Grumman F-14 Tomcat fighters operated by the US Navy, the similar missiles and aircraft supplied to the Imperial Iranian Air Force being no longer serviceable. The missile was designed in parallel with the Hughes AWG-9 radar fire-control system for installation in the General Dynamics F-111B naval fighter, and when the overweight F-111B was cancelled became the basis for Grumman's successor design. The AWG-9 radar has a look-down search range of 150 miles/241 km or more and can track multiple targets, and the Phoenix was optimized to match the radar's capability.

Weighing 985 lb/447 kg at launch, the Phoenix has a speed in excess of Mach 5 and a range in excess of 125 miles/201 km, performance provided by a long-burning Aerojet Mk 60 or Rocketdyne Mk 47 solid-propellent rocket. After launch the missile climbs to about 100,000 ft/30,480 m and attains a burn-out speed of Mach 3.8 under the control of an autopilot which steers the missile towards the anticipated target position. After rocket burn-out, the missile slowly descends and accelerates to Mach 5+: as it does so, it acquires the target using its radar in the semi-active mode to receive electromagnetic energy bounced off the target by the launch aircraft's radar.

As the missile homes on the target its radar switches to active mode, transmitting its own signal and using its antenna to determine the target's exact position and range. For its size the Phoenix is remarkably agile even at long range, since the control fins aft of its delta wings are hydraulically powered. The warhead is a devastating 132 lb/60 kg unit carried just ahead of the wings; it is initiated either on impact or at the optimum distance from the target by a Downey Mk 334 radar or Bendix infra-red proximity fuse.

The initial AIM-54A variant introduced in 1973 was superseded in 1977 by the AIM-54B, which has a simplified structure, wings of sheet metal rather than honeycomb and non-liquid hydraulic and cooling systems. Development is currently under way of the AIM-54C model: matched to the latest Tomcat variant, this features an upgraded Nortronics interial reference unit, improved electronic counter-countermeasures and solid-state digital electronics in place of the earlier variants' klystron-tube analog electronics.

In the late 1960s the trend began to swing the other way with the semi-recessed or palletized carriage of missiles on aircraft such as the F-14 Tomcat, and similar arrangements are evident on the new generation of combat aircraft, especially those with an air-combat and/or air-superiority role for which minimization of drag is very important. This tendency is certain to continue, and may lead to the reintroduction of internal weapons bays for small air-to-air missiles carried in large numbers.

CORE AIRCRAFT

The process of treating the fuselage and wing roots as a structural core to which can be attached an evolutionary series of flying surfaces has been evaluated by the Rockwell HiMAT (Highly Manoeuvrable Aircraft Technology) remotely piloted vehicle of the early 1980s. The programme failed to explore the full range of possibilities, but the scope of the original concept emphasizes the way evolutionary development of core aircraft could proceed. Using the core as the structural, electronic and propulsive foundation, the HiMAT was flown in a canard layout and was schemed with a planar arrow wing, variable-incidence wing and forward-swept wing, and was also envisaged with a two-dimensional vectoring-thrust nozzle on its engine.

Evolutionary design of this type makes excellent economic as well as aerodynamic sense. Before the mid-1950s the peacetime service life of a combat aircraft could be measured in months, and was far less in war. Since then the cost of developing and manufacturing each type has escalated so enormously that the designer has to consider a service life of 15 or more years before a completely new design can be perfected for service and actually afforded by the operator arm. It is impossible for even the most skilled design team to predict the range and type of technological improvements that may become available in that time, yet it is essential that such improvements should be incorporated whenever possible to prevent a potential enemy from securing any decisive technological advantage.

OPPOSITE, ABOVE The Rockwell HiMAT is one of the most fascinating of modern research tools into aerodynamic and aircraft structures. It is a remotely-controlled miniature aircraft capable of supersonic performance. The modular design allows the machine to be configured in several ways on a standard core section.

OPPOSITE, BELOW The HiMAT in flight.

BELOW The HiMAT has allowed the investigation of flight at very high angles of attack, with a close-coupled canard configuration.

GRUMMAN F-14A TOMCAT

TYPE: two-seat carrier-borne fleet air-defence fighter with variable-geometry wings
WEIGHTS: empty 40,104 lb/18,191 kg; maximum take-off 74,348 lb/33,724 kg
DIMENSIONS: span 64 ft 1½ in/19.54 m spread and 38 ft 2½ in/11.65 m swept; length 62 ft 8 in/19.10 m; height 16 ft 0 in/4.88 m; wing area 565 sq ft/52.49 m² spread
POWERPLANT: two 20,900-lb/9,480-kg afterburning thrust Pratt & Whitney TF30-PW-412A turbofans
PERFORMANCE: speed 1,545 mph/2,486 km/h; ceiling 50,000+ ft/15,240+ m; range 2,000 miles/3,220 km
ARMAMENT: one 20-mm multi-barrel cannon and 14,500 lb/6,577 kg of disposable stores (generally four Phoenix long-range, two Sparrow medium-range and four Sidewinder short-range air-to-air missiles) on four underfuselage pallets and two underwing hardpoints

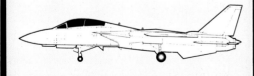

ABOVE The Grumman F-14A Tomcat has been dogged throughout its life by powerplant deficiencies, but is still a type that attracts superlatives in the description of its overall design. This example carries two AIM-7 Sparrow and two AIM-9 Sidewinder missiles under the outer ends of its fixed centre section, leaving the underfuselage positions possibly occupied by four more Sparrows or more probably a quartet of AIM-54 Phoenix long-range missiles.

LEFT The F-14s Tomcat, foreshadowed in this Super Tomcat prototype, introduced digital electronics and other improvements, plus a considerably improved powerplant for greater flexibility of operation.

RIGHT The Tomcat in its operational regime, providing US Navy carriers with long-range defence against air and missile attack.

ABOVE The large size of the AIM-54 Phoenix is readily apparent as the deck crew of USS *America* prepares a Tomcat for a sortie.

RIGHT The Tomcat has maximum flexibility of air superiority firepower, as revealed on this example which carries two short-range AIM-9 Sidewinders, two medium-range AIM-7 Sparrows and at least two long-range AIM-54 Phoenixes. There is also a 20-mm Vulcan cannon in the port side of the forward fuselage.

ABOVE RIGHT Tomcats on the flight deck of USS *Constellation* with their wings fully swept to reduce deck area requirement.

Consequently, the overall thrust of aircraft design in recent years has switched away from the concept of the aircraft as a throw-away item to be replaced as soon as a new model becomes available, toward the notion of the core airframe and powerplant as the basis for continuing development. Already, combat aircraft are designed with a mid-life update in mind, using modular features and databuses to allow the installation of more advanced components. The F-14D version of the Tomcat, for example, will be an altogether more formidable warplane than the current F-14A. New-generation digital electronics will replace the original analog electronics, upgraded weapons will be matched to the new electronics, and new engines will provide 30% more power plus far greater flexibility and reliability.

NEW MATERIALS

Aerodynamics are ineluctably tied to structure: it is good news when the design team's aerodynamicist comes up with the right answer to a requirement, but unless the structural designer can turn this answer into hardware it is of no practical value. Wood was the first primary structural medium for aircraft, but it rapidly faded from use in advanced aircraft as aluminium alloys came to the fore. Such alloys had been used experimentally even before World War I, together with steels of various types, but from the early 1920s the advantages of aluminium alloy as the primary structural medium dictated the development of powered aircraft.

At first aluminium alloys were used merely as a substitute for wood, aluminium alloy tubes simply replacing wooden longerons and spacers in a primary structure that was internally braced with wires and then covered with fabric or plywood. Then the implications of semi-monocoque construction were fully digested and the designers realized that an aluminium alloy skinning, suitably reinforced on its inner side by stringers and frames, could be used for combined aerodynamic and structural purposes. The resulting elimination of much of the internal structure and bracing that had increased structural weight and reduced usable internal volume, as well as making it difficult to produce effective cantilever structures, paved the way for the modern aircraft with its markedly enhanced performance and features such as retractable landing gear, high-lift devices and enclosed cockpits that have become standard.

Aluminium alloys continued to dominate the structural scene until the early 1980s. There have been exceptions, but mainstream aircraft production remained faithful to aluminium alloys of one kind or another up to the point at

GEODETIC CONSTRUCTION

Though no longer used, the geodetic system of construction enjoyed considerable popularity in the 1930s and early 1940s. Developed largely by Sir Barnes Wallis for a series of airships built by Vickers it takes its name from the geodetic line, the shortest line between two points on a surface.

Geodetic construction involves the creation of curved space frames whose individual members follow geodetic lines along the surface, each undergoing either compression or tension. The resultant structure looks somewhat like open-weave basketware, but has the advantages of great strength at modest weight. As well as creating a truly fail-safe structure it obviates the need for a stressed-skin covering and, in the context of the 1930s, was relatively straightforward to build.

The structure of individual members was pioneered in the disastrous M.1/30 biplane torpedo bomber and a primitive geodetic structure was used for the G.4/31, another biplane bomber. The latter had four light-alloy longerons over which were wound clockwise and anti-clockwise spiral channels to create a light yet very strong lattice structure.

The first aeroplane with a proper geodetic structure was the Wellesley monoplane bomber, the type used to set a world distance record of 7,158.444 miles/11,520.42 km between Ismailya in Egypt and Darwin in Northern Australia between November 5 and November 7, 1938. But the ultimate expression of the geodetic structure was the Wellington bomber of World War II.

In the Wellington Wallis refined his concepts into an immensely strong structure that ultimately proved simple to make once the semi-skilled work force had grown accustomed to the required techniques. Subsequent Vickers bombers built with a geodetic structure were the modestly successful Warwick, designed as a heavier counterpart to the Wellington, and the unsuccessful Windsor four-engined heavy bomber.

which aerodynamic heating began to enter the picture. Other materials were used in lightly-loaded areas, and glass-reinforced plastics (GRP) became particularly popular. GRP is used for complex shapes where light weight was desired, such as wingtips and fairings in areas of minimal aerodynamic tension, and for coverings over electronic installations such as radomes.

FIBRE-REINFORCED COMPOSITES

From GRP has been developed a new structural medium that may well prove as significant as aluminium alloys in the evolution of aviation technology. This is fibre-reinforced composite (FRC), which comes in a variety of forms that can be tailored to exact and exacting technical specifications. FRC is essentially GRP further reinforced with a layer or layers of lightweight carbon-graphite or boron fibres of exceptional stiffness and strength. Unlike a metal sheet of uniform thickness, which flexes equally in two planes, FRC can be made undirectionally stiff: the fibres can be laid in such a fashion that the resultant sheet of FRC material will flex in one plane without difficulty, but remain rigid in the other plane.

FORWARD-SWEPT WINGS

FRC materials have opened the way for the forward-swept wing, whose advantages have long been recognized by aerodynamicists. The first such wing was flown as far back as 1944 in the Junkers Ju 287 research aircraft intended as a prototype bomber. As the Germans discovered, however, the structural materials of the day could not prevent the phenomenon of aeroelastic divergence: any sudden control movement can cause the wings to flex, causing one wing's angle of incidence to increase, thereby generating additional lift (structural load) and yet more divergence (steadily increasing structural load) in a rapid cycle that results in the wing being torn off in a few fractions of a second by uncontrollable aerodynamic forces. Swept-back wings do not suffer from this structural problem, since the load rapidly reduces and damps rather than increases the aeroelastic divergence.

Nevertheless, the advantages of swept-forward wings are theoretically enormous. They delay any rise in transonic drag, reduce stalling speed, enhance low-speed handling

ABOVE AND LEFT The de Havilland Mosquito was an early exponent of composite construction, though in this instance the composite was a sandwich of balsa between outer layers of plywood. The result was a smooth, tough and easily repairable structure that was light and made minimum demands on the strategic alloys that were in short supply during World War II.

characteristics and provide virtually stall-proof flight. In combination these features should make it possible for a smaller aeroplane to have greater performance and agility on a given level of power, or the same performance plus greater agility on less power. The advantages in economic and tactical terms are enormous, and the opportunity offered by FRC has been widely explored, principally by the Grumman X-29.

The X-29, which first flew in December 1984, is purely a research tool, and uses the forward fuselage and cockpit of a Northrop F-5 fighter to reduce development costs. A design of relaxed static stability, it is powered by a General Electric F404 afterburning turbojet for a maximum speed of Mach 1.6+. The type's *raison d'être* is the forward-swept wing structure, which is attached about two-thirds of the way aft along the fuselage.

OPPOSITE, TOP A piece of skinning is lowered onto the wing of an F-15 fighter under construction by McDonnell Douglas. The panel is of a new aluminium-lithium alloy, offering comparable strength but lighter weight than conventional aluminium alloys.

OPPOSITE, BELOW The Grumman X-29 is an invaluable research aeroplane, pioneering the flight evaluation of a forward-swept wing.

RIGHT, BELOW AND BOTTOM Only the development of carbon-based composite materials has made possible the development of the X-29's forward-swept wing, which offers aerodynamic advantages over the aft-swept wing – but only if the structure can be tailored to avoid twisting.

The core of the wing structure is an electron-beam-welded box of titanium and light alloy, providing an exceptionally sturdy but generally conventional basis for the outer aerodynamic surfaces. The latter are single-piece upper and lower skins made of CFRP (carbon-fibre-reinforced plastics) up to 156 layers thick at their inboard ends. The skins are exceptionally light yet rigid, and can sustain the most violent aerial movements without any possibility of aeroelastic divergence. The leading edges are fixed, with no provision for high-lift devices of any kind, but the trailing edges are fitted with full-span flaperons that can be used as camber-changing sections. Flight tests have confirmed that these provide the forward-swept wing with very nearly the same mission capabilities as the Boeing MAW, with the added advantages of forward sweep and reduced mechanical complexity.

Located aft of the wing are the conventional rudder, plus a pair of strake flaps fitted at the extreme rear of the extended wing root trailing edges, nearly in line with the rudder. Powerful canard foreplanes with one-fifth of the wings' area are located on the sides of the lateral inlets, and just forward of the inboard sections of the wing leading edge, which are conventionally swept back. The canards are driven though a triply-redundant fly-by-wire flight-control system, and are the aeroplane's primary con-

LEFT The McDonnell Douglas AH-64A Apache is an inelegant but magnificent helicopter with good performance and highly capable electronics.

BELOW Complete with 16 missiles and underfuselage cannon, the Apache has a pugnacious appearance emphasized by the nose-mounted assembly of essential sensors.

RIGHT The attack helicopter is becoming increasingly important for the more sophisticated carriers, and the Apache is a leading contender in this field with the weapons displayed here: 30-mm rounds for the underfuselage cannon, 2.75-in/70-mm rockets for the multi-tube launchers, and up to 16 laser-homing AGM-114 Hellfire anti-tank missiles

trol surfaces in the pitching plane. In common with other canard surfaces, they are used to trim out any tail-up pitching moment by generating lift, augmenting the lift of the wing rather than reducing it in the fashion of tail-mounted pitch-control surfaces. The relationship of the canards and wings is mutually beneficial: the canards gain from an effective doubling of their moment arm, and the wings benefit from the downwash from the canards.

Flight tests have confirmed wind-tunnel predictions about the X-29's flight characteristics: even at extremely high angles of attack the aircraft cannot be stalled and it retains full roll authority down to very low airspeeds. Early flight trials also indicated that fuel burn was lower than expected, an indication of exceptionally low drag. There is every reason to forecast that the swept-forward wing made possible by FRC will become the norm for air-combat fighters, which will profit enormously from its increased agility. Attack aircraft should also benefit from the reduction in drag offered by the configuration.

Despite the success of the X-29 programme, current attention seems fixated on variable-camber wings as the primary means of optimizing the payload/range equation. There is no doubt that the MAW concept has great advantages over the conventional wing, but that is only achieved at the cost of mechanical complexity. Such complexity may be acceptable in a civil airliner, but is surely fraught with dangers in a military type, which must expect far more rigorous handling and the possibility of critical battle damage.

THE WAY FORWARD

FRC construction is certainly one of the most important ways forward in aircraft design, providing not just greater strength at reduced weight, but the possibility of tailoring that strength to secure specific aerodynamic advantages. As the technology matures, FRC is being used to replace aluminium alloys in new-generation combat and civil aircraft. In the longer term there is every reason to believe that more sophisticated FRCs will cause the very nature of aerodynamic design to be revised.

At the same time, there are new alloys such as aluminium-lithium, now being produced in bulk in the US, which will find extensive applications in civil and military aircraft.

Recently, doubts have been expressed about the ability of FRC to cope with large-calibre cannon projectiles, but the manufacturers claim that the problems have been exaggerated and will be overcome as FRC matures in development and use. The largest FRC structure in military service is the wing of the Harrier II, a super-critical structure built of graphite/epoxy composite as a single unit.

HELICOPTERS

The helicopter differs from almost all fixed-wing aircraft in its ability to take off and land vertically and hover in the air. And while helicopters will never be able to match fixed-wing aircraft in any outright parameter of performance, their ability to translate at will between vertical and horizontal flight means that they can operate from confined areas such as jungle clearings and the heaving decks of small warships.

Helicopters began to appear in operational service in the last two years of World War II, but the value of early examples was limited by their marginal performance. The breakthrough in helicopter capability came with the turboshaft powerplant. Turboshafts are both smaller and more powerful than piston engines; they can be located close to the rotor assembly, simplifying and lightening the associated transmission system and leaving a clear volume for payload. Other advantages are greater power in overall terms, increased reliability and significantly reduced vibration levels.

It was the French who pioneered turboshaft-powered helicopters, followed closely by the Americans and then the Soviets. In its new guise the helicopter became a useful flying machine and an invaluable tool on the modern battlefield, rapidly evolving from the utility machine epitomized by the Bell UH-1 Huey into specialized types such as the light scouting helicopter, exemplified by the Hughes OH-6 Cayuse and dedicated gunships like the Bell AH-1 HueyCobra.

Trainable nose turrets for grenade-launchers, multi-barrel cannon or machine guns and hardpoints on stub wings for gun pods, rocket launchers and air-launched anti-tank missiles have become standard features on battlefield helicopters, as their potential has been realized.

At the other end of the scale are medium- and heavy-lift helicopters such as the twin-rotor Boeing Vertol CH-47 Chinook, the single-rotor Sikorsky CH-53 Sea Stallion series and the Mil Mi-26 Halo. These large machines are used to move troops and equipment on the battlefield and in support of amphibious forces.

The naval helicopter, meanwhile, has become a multi-role machine whose most important task – the detection and attack of submarines – is performed by such types as the Sikorsky/Westland Sea King and the Sikorsky SH-60 Seahawk. Naval helicopters can also attack small surface vessels using missiles such as the Kongsberg Penguin, British Aerospace Sea Skua and Sistel Sea Killer.

The growing electronic capability of the larger helicopters and their greater weight-carrying ability allow them to carry heavyweight anti-ship missiles such as the British Aerospace Sea Eagle and Aérospatiale Exocet, designed for use against large warships.

Developments are continuing apace, and modern helicopters such as the Westland Lynx combine a useful payload with excellent flight performance and aerial agility. The result is an increasing shift in emphasis toward fighter helicopters. Air-to-air versions of shoulder-launched surface-to-air missiles provide them with the ability to tackle heavier machines such as the Mil Mi-24 Hind and McDonnell Douglas AH-64 Apache attack helicopters.

ABOVE The composite wing of the McDonnell Douglas AV-8B Harrier II is light, large, and strong enough to support six underwing hardpoints for this versatile machine's wide array of armament options.

LEFT AV-8B construction, with the forward fuselage about to be mated to the aft fuselage before the installation of the vectored-thrust turbofan and the one-piece wing.

BELOW A ghosted illustration reveals some of the Harrier II's primary features: the black boxes contain electronics, the pale grey/blue colour shows the powerplant and thrust-vectoring system, the red highlights the fuel system, the middle blue reveals the reaction control system for control in the hover, the pale green marks the auxiliary power unit and the yellow is used for the air-conditioner/heat exchanger.

RIGHT The AV-8A is the US Marine Corps' equivalent to the British Harrier GR Mk 3, with the original small wing.

LEFT AND BELOW RIGHT The AV-8B: note the large number of design differences between this and the earlier AV-8A type.

2
POWERPLANT

All modern combat aircraft are powered by turbine engines of the type pioneered in combat during 1944 by the Gloster Meteor and Messerschmitt Me 262. The turbojet, universal until the mid-1960s, is a comparatively simple engine comprising compressor, combustion chamber and turbine, the last extracting just enough power from the gas flow from the combustion chamber to drive the compressor. Most of the power remains in the gas flow, which is expanded into the atmosphere at high velocity through a constricting nozzle and is often augmented by the afterburner. The afterburner – or reheat, as it is known in Britain – injects additional fuel into the exhaust: this fuel mixes with the free oxygen remaining in the exhaust gas and burns to create additional gas exhausted through a nozzle of larger area. The afterburner is effective only in supersonic flight; it can boost basic thrust by as much as two-thirds, but only by consuming disproportionate amounts of fuel.

The turbojet is still used for aircraft optimized for a single role or flight regime, whether subsonic or supersonic, though it is significant that aircraft powered by afterburning turbojets are those, like the MiG-25 interceptor and SR-71 reconnaissance aircraft, which are designed for

OPPOSITE, ABOVE A good example of how the powerplant conditions overall design is provided by this English Electric Lightning F Mk 6, whose fuselage accommodates a vertical pair of Rolls-Royce Avon afterburning turbojets, fuel and, large circular inlet.

OPPOSITE, BELOW In the years immediately after World War II airline operations were dominated by the USA's huge air-cooled radials, which offered great power, efficiency and reliability. This Lockheed L-749 Constellation has four Wright Double Cyclones.

ABOVE US Navy fighters such as this Boeing F4B-4 were largely responsible for the early development of the radial engine, in this case the Pratt & Whitney R-1340.

LEFT The Boeing 247 was the first 'modern' airliner, and was powered by two Pratt & Whitney Wasp radials.

MESSERSCHMITT Me 262 AND JUNKERS JUMO 109-004

The most significant combination of airframe and powerplant during World War II was that of the Messerschmitt Me 262 and its Junkers Jumo 109-004 turbojet. Junkers had started work on turbojets as early as 1936 and opted immediately for the axial-flow type rather than the less technically demanding centrifugal-flow type. The first Jumo 109-009 was running by the autumn of 1938, but the project and its staff were switched to Heinkel after Junkers' engine division discovered that jet engine development work was being undertaken in secret under the auspices of the company's airframe division.

The engine division of Junkers had already begun to move into turbojet development, and in the summer of 1939 it received a German air ministry contract for the new Jumo 109-004. This was designed for a thrust of 1,543 lb/700 kg at a speed of 559 mph/900 km/h, and was schemed round an eight-stage axial compressor, six combustors, a single-stage turbine designed with the aid of AEG's turbine expertise and provision for afterburning.

The first unit ran in November 1940, but was beset by so many technical problems that it was early 1942 before anything approaching reliability was attained; a Jumo 109-004A was flown for the first time under a Messerschmitt Bf 110 in March 1942, and other A-series engines were flown in prototypes of the Me 262. In the production-configured B-series the quantity of strategic materials was halved, largely through the replacement of castings by sheet metal constructions. This reduced both weight and the number of man hours involved in building each engine.

The Jumo 109-004B-1 first ran in May 1943, and the use of blades of improved shape in the first two stages of the compressor helped to increase thrust usefully to 1,984 lb/900 kg. This variant was quite rapidly cleared for production, the first units being delivered in March 1944 to power the classic Me 262A. Further development resulted in the Jumo 109-004B-4, with hollow rather than solid turbine blades, which entered production in December 1944; it was about to be supplanted by the 2,315-lb/1,050-kg thrust Jumo 109-004D-4 as the war ended. The Junkers plants and their work forces were seized by the advancing Soviets, and the whole production facility was moved to the USSR for continued development.

straight-line Mach-3 performance at high altitudes. Smaller turbojets are still used to power drones, remotely-piloted vehicles and sea-skimming anti-ship missiles.

THE TURBOFAN

The most common powerplant for modern aircraft, both civil and military, is the turbofan. This has the same basic core as the turbojet, but uses a far higher proportion of the gas flow from the combustion stage to drive the turbine: the turbine has more stages and drives a fan that both compresses the air for the core stage and ducts a substantial mass of propulsive air around the core section to generate the bulk of the thrust. For a given fuel consumption such an engine generates far greater thrust than a turbojet, especially at lower airspeeds, and is much quieter. The reduction in noise results from the cylinder of cooler air driven by the fan, which surrounds, smooths out and finally cools the hot and highly turbulent gas flow from the combustion chamber. The cooler air also reduces the engine's infra-red signature which forms the principal target for heat-seeking missiles. Like the turbojet, the turbofan can be augmented by an afterburner operating in the hot or cold gas flows.

The turbine has come a long way since World War II in terms of power output and reliability, while shrinking in weight and size and improving its specific fuel consumption – the unit of fuel required to generate a specific unit of thrust for a given period of time. For example, the Allison J35-A-35 of 1951 was 27 in/0.686 m wide and 195.5 in/4.97 m long, weighed 2,850 lb/1,292.76 kg, and produced an afterburning thrust of 7,500 lb/3,402 kg at a specific fuel consumption of 2.0. By 1965 matters had improved dramatically, the General Electric J79-GE-17 being 39.06 in/0.99 m wide and 208.7 in/5.30 m long, weighing 3,847 lb/1,745 kg, and delivering an afterburning thrust of 17,820 lb/8,083 kg at a specific fuel consumption of 1.97. Yet this achievement, producing one of the most widely used of all turbojet engines, is eclipsed by that of the Turbo-Union RB199 used to power the Tornado: only 34.25 in/0.87 m wide and just 127 in/3.225 m long, the RB199 weighs a mere 1,980 lb/898.1 kg while delivering an afterburning thrust of 16,000 lb/7,257.6 kg) with a specific fuel consumption of 1.5.

ABOVE AND LEFT The almost incredible Lockheed SR-71A 'Blackbird' strategic reconnaissance platform is powered by a pair of Pratt & Whitney J58 (JT11D-20B) bleed turbojets burning a unique fuel. These engines each develop a dry thrust of 23,000lb/10,430kg rising to 32,500lb/14,740kg with afterburning, and at high speeds develop much of their power as suction at the inlet.

McDONNELL DOUGLAS AGM-84 HARPOON PROPULSION

Missiles are generally associated with rocket motors of the solid-propellant type, but in recent years there has been a significant switch away from such propulsion systems. The reason is simple: solid rockets must carry both fuel and the oxidizer required to make it burn, limiting both the quantity of fuel and the missile's range. This hardly matters in short-range weapons such as dogfighting air-to-air missiles, for which rapid acceleration and instant power are more important. But where range is a principal requirement, as it is for sea-skimming anti-ship missiles, it is a severe handicap.

Sea-skimmers fly in the oxygen-rich lower reaches of the atmosphere, where it seemed pointless for them to carry their own oxidizer. The logical result was a new generation of subsonic missiles powered by small turbines: truly long-range weapons such as the Boeing AGM-86 air-launched cruise missile, with diminutive turbofans, and shorter-ranged weapons such as the AGM-84 Harpoon with one-shot turbojets.

The Harpoon's Teledyne CAE J402-400 evolved ultimately from the French Turboméca Marbore II, which was Americanized as the J69 as the start of Teledyne's mini-turbojet development programme in 1951. In the AGM-84 the J402 is rated at a sea-level thrust of 660 lb/299 kg turning at 41,200 revolutions per minute. A trim unit with precision-cast axial- and centrifugal-flow stages, it is located in the tail. Exhausting centrally between the rear control fins, it is fed with air via a ventral inlet and with fuel from a tank located above the inlet on the missile's centre of gravity, thereby avoiding shifts in the centre of gravity as the fuel is consumed.

FAIRCHILD REPUBLIC A-10A THUNDERBOLT II PROPULSION

One of the most interesting propulsion arrangements on a current combat aircraft is that of the Fairchild Republic A-10A Thunderbolt II. Designed for very modest overall performance, but requiring the ability to loiter for long periods over the battlefield, the A-10 relies on a pair of turbofans to produce the required endurance without excessive fuel consumption.

The engine involved, the General Electric TF34, was developed for the Lockheed S-3 anti-submarine aircraft, which also needs to be able to loiter for long periods at low levels. This useful engine has a single-stage fan and a 14-stage axial-flow compressor, is rated at 9,275-lb/4,207-kg thrust and weighs 1,478 lb/670 kg.

Just as important in the evolution of the A-10A was the engines' location on short pylons extending above and slightly out from the fuselage between the wings and tailplane. Such an installation is satisfactory from the aerodynamic point of view, offering minimal control problems in the event of loss of a single engine. More importantly, it means the wings, tailplane and much of the fuselage are between the ground and the engines, shielding the engines from ground fire and hiding them from ground-based sensors.

The flat face of the fan is the feature of turbofan-powered aircraft most readily detected by radar, and the A-10A's wing is below and forward of the engines to protect them from such detection. At the other end of the engines the tailplane conceals the exhaust plumes, whose signature is already reduced by the cold air propelled by the fan, from the gaze of infra-red homing missiles.

FAIRCHILD REPUBLIC A-10A THUNDERBOLT II

TYPE: single-seat battlefield close-support and anti-tank aircraft
WEIGHTS: empty 24,960 lb/11,322 kg; maximum take-off 50,000 lb/22,680 kg
DIMENSIONS: span 57 ft 6 in/17.53 m; length 53 ft 4 in/16.26 m; height 14 ft 8 in/4.47 m; wing area 506 sq ft/47.01 m^2
POWERPLANT: two 9,065-lb/4,111-kg thrust General Electric TF34-GE-100 non-afterburning turbofans
PERFORMANCE: speed 439 mph/706 km/h; ceiling not relevant; range 576 miles/927 km with a 1.7-hour battlefield loiter
ARMAMENT: one 30-mm Avenger multi-barrel cannon and up to 16,000 lb/7,258 kg of disposable stores (including two or four Sidewinder short-range air-to-air missiles, six Maverick air-to-surface missiles, glide bombs with laser or electro-optical guidance, cannon pods, rocket pods and conventional or cluster bombs) on three underfuselage and eight underwing hardpoints

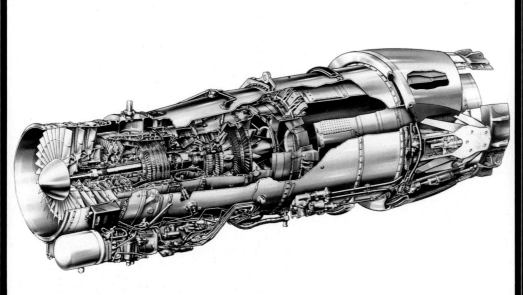

ABOVE The Tornado ADV's massive inlets provide striking evidence of the air quantities gulped by modern turbofans.

LEFT The Turbo-Union RB199 Mk 104 is a classic example of high power from an extremely compact design. Note the clamshell thrust-reverser round the nozzle.

ABOVE RIGHT Optimization of the RB199's thrust requires the use of fully-variable nozzles for the two engines.

ABOVE FAR RIGHT The Tornado ADV is an immensely impressive sight in full afterburner.

RIGHT The RB199 Mk 104D is used in the British Aerospace EAP research fighter.

BYPASS RATIOS

Turbofans are generally assessed in terms of their bypass ratios – that is, the ratio between the cool air accelerated by the fan past the core section and the high-pressure air used in the core section. There is a great diversity between the turbofans used in combat aircraft and those that power transports. Combat aircraft must use an engine of small diameter, and this militates against the use of a large fan.

Typical turbofans used by current combat aircraft are the General Electric F404 (bypass ratio 0.34), Prat & Whitney F100 (0.7), Turboméca/Rolls-Royce Adour (0.8), Turbo-Union RB199 (1) and Rolls-Royce Pegasus (1.4). The higher the bypass ratio the better the specific fuel consumption is a general rule, and modern combat aircraft with high-bypass-ratio turbofans have significantly longer ranges than earlier aircraft on the same fuel load.

It is interesting to look at the evolution of the F-16's powerplant in the days when it was still the General Dynamics Model 401. Some thought was given to a powerplant of two General Electric J101 afterburning turbojets, which gave a mission weight – airframe, fuel and weapons – of 21,470 lb/9,739 kg. Then the concept was re-cast with a single Pratt & Whitney F100 afterburning turbofan, whereupon mission weight declined to 17,050 lb/7,734 kg, offering various useful possibilities. To give just one, the weight saving could be used for additional fuel, boosting range by more than 70%.

The range increment was attributable to the lighter powerplant installation (45%), lower fuel flow (40%), reduced airframe weight (11%) and reduced drag (4%).

Turbofans for fighters inevitably involve a compromise in design emphasis and effort, being designed to provide aircraft with both high performance and long range, as well as being optimized for extreme flexibility of operation so that the pilot can work the throttle as rapidly as required by the tactical situation. The high-bypass-ratio turbofan for long-range cruising performance is entirely different. Such engines are used in modern airliners and long-range military transports. Immediately noticeable for the diameter of their enormous fans, they have bypass ratios in excess of 4 allied to extremely good fuel consumption.

LEFT The nature of the Rolls-Royce Pegasus non-afterburning turbofan is readily apparent from this view showing the two cold-gas forward, and two hot-gas aft, thrust-vectoring nozzles.

BELOW LEFT Finnish technicians check a Turbomeca/Rolls-Royce Adour non-afterburning turbofan before its installation in a BAe Hawk trainer. The Adour has been developed in afterburning and non-afterburning forms, both notable for their good power-to-weight ratios.

RIGHT Clearly visible on this F-16 two-seater is the fully-variable nozzle for the Pratt & Whitney F100 afterburning turbofan, aspirated via an inlet under the forward fuselage to optimize airflow at high angles of attack.

BELOW A ghosted view of the two-shaft Adour shows (from front to rear) the two-stage fan, five-stage high-pressure compressor, annular combustion chamber, single-stage high-pressure turbine, single-stage low-pressure turbine and nozzle.

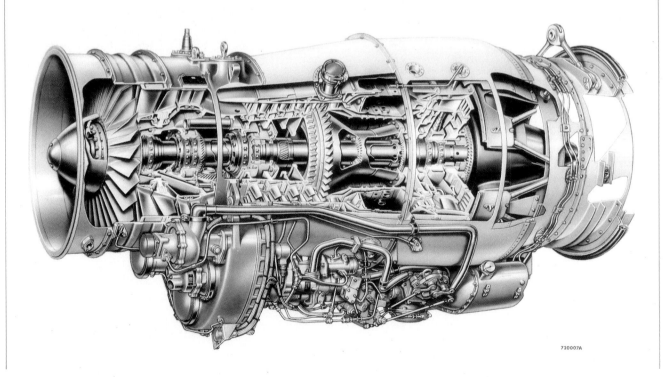

RIGHT Like the F-16, the McDonnell Douglas F/A-18A Hornet is optimized for modestly supersonic performance, though in this instance with a side-by-side pair of General Electric F404 afterburning turbofans.

BELOW LEFT The powerplant of the F/A-18A was designed for optimum operating capability and easy maintenance, the two F404s being located in 'straight-through' installations on the sides of the lower fuselage.

BELOW RIGHT Through retaining the standard F100 engine of its F-16B origins, the F-16XL has much enlarged fuel capacity.

OPPOSITE, ABOVE LEFT The F/A-18A has simple inlets, perfectly adequate for the type's modest performance.

OPPOSITE, BELOW The F/A-18A's powerplant is concentrated in the rear of the airframe.

OPPOSITE, ABOVE RIGHT The powerplant of the F/A-18A, like that of other modern aircraft, is voluminous: in red are the engines, and in yellow the fuel tanks.

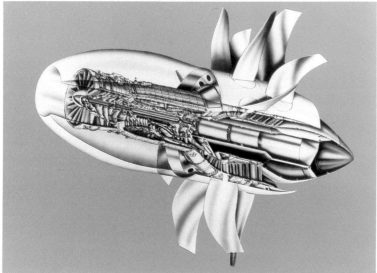

THE UNDUCTED FAN

A development of the late 1980s that seems set to complement the high-bypass-ratio turbofan is the unducted fan (UDF): this can be regarded as the modern equivalent of the turboprop, in which the main power of the turbine is used to drive a substantial propeller, but combines the turbofan's performance and fuel economy with the turboprop's low noise.

A characteristic of such engines, examples of which are under development in the USA by Allison, General Electric and Pratt & Whitney, is the use of an aft-mounted fan with multiple blades, made of aeroelastically stiff FRC and of swept configuration to mitigate the onset of compressibility problems as the tips approach Mach 1. Fears about the damage that might be caused by a blade separating from its fan mean that such engines are most likely to be tail-mounted. Most of the applications currently envisaged involve medium-capacity civil aircraft, but once the concept has been fully validated it is reasonable to expect that the UDF will be used for long-range miliary aircraft.

INLETS AND NOZZLES

Just as important as the specific engine is the arrangement of its inlets and nozzles. The nozzle is the region of the engine at which the hot gas flow is discharged into the atmosphere, its role being to convert as much as possible of the total energy of the gas into kinetic energy. The nozzle generally has a greater diameter than the engine itself, and is most commonly of the convergent/divergent type, in which the cross section converges into a throat where the subsonic gas flow is accelerated to supersonic speed, then diverges in order to allow further acceleration of the gas flow as it emerges into the atmosphere.

The convergent/divergent nozzle is fully variable to ensure that optimum conditions can be provided in all flight regimes, being constructed in a series of overlapping petals controlled by a powerful actuator system.

The design of the nozzle area is critical to the operation of the whole engine, and presents extreme aerodynamic difficulties as well as the expected metallurgical problems associated with an incandescent and vibration-charged plume of gas emerging from the engine.

The thrust of the engine can be used for

IN-FLIGHT REFUELLING

In-flight refuelling is such an integral part of air operations that it is easy to forget that the practice is a comparatively recent innovation for all but heavy bombers. Its most obvious use is to top up an attacking aircraft's tanks before it enters enemy airspace; but it can also enable aircraft to carry heavier loads by trading fuel for payload during takeoff and to sustain damaged aircraft.

There are now two basic techniques for in-flight refuelling, one using a flying boom, the other known as the probe-and-drogue type. The former involves a rigid boom controlled from the tanker: an operator uses the boom's aerodynamic controls to position its tip in the receptacle on the upper surface of the receiving aircraft, which formates precisely on a system of markings and lights on the tanker's underside. The flying boom system can be carried only by large aircraft with special tankage for rapid fuel transfer.

The hose-and-drogue system offers greater tactical flexibility. The tanker, which can be a combat aircraft fitted with a buddy refuelling pack drawing fuel from its own fuel system, trails a flexible hose fitted with a stabilizing drogue containing the female portion of the fuel-transfer coupling: the receiving aircraft flies into position below and aft of the tanker, and inserts its refuelling probe.

braking as well as acceleration. Most civil aircraft are fitted with a system of thrust reversal, normally in the form of cascades or clamshells to divert the thrust forward and so exert a powerful decelerative force.

Increasingly, however, the virtues of dispersed-site or damaged-runway operations are persuading the military powers that thrust reversal offers persuasive tactical advantages.

The most notable combat aircraft to incorporate thrust reversers are the Tornado and the Viggen, both exceptionally potent types: that of the Tornado is fitted upstream of the fully variable nozzle, and that of the Viggen allows the aeroplane to be flown straight onto the ground without a flare; as soon as the Viggen's nosewheel touches the ground and its leg is compressed the thrust-reversal system is automatically activated to bring the aircraft to a rapid halt.

OPPOSITE, TOP The General Electric UDF engine on test in the starboard position of a Boeing 727.

OPPOSITE, MIDDLE The UDF engine combines the best features of the turbofan and turboprop.

OPPOSITE, BOTTOM The Pratt & Whitney JT9D-7R4 is one of the world's most economical turbofans and is used on the Airbus A300-600 and A310, and on the Boeing 747SUD and 767.

ABOVE RIGHT Still one of the most impressive powerplants in service, the Rolls-Royce/SNECMA Olympus 593 afterburning turbojet is batched in pairs under the wings of the Concorde supersonic airliner.

RIGHT The Rolls-Royce offering for large civil transports is the RB211, seen here in the form of the RB211-535C variant.

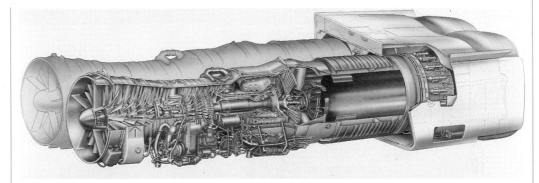

ABOVE The use of pylon-mounted pods for the engines of major airliners allows the purchasing airline to choose between several possible engines, the British Airways Boeing 757 being powered by a pair of Rolls-Royce RB211-535C turbofans.

LEFT Standard powerplant on the BAe 146 is a quartet of Avco Lycoming ALF 502R-5 turbofans, each rated at a thrust of 6,970lb/3,160kg. The use of four small turbofans increases reliability and reduces noise levels, allowing the BAe to operate by night into and out of airports otherwise closed to turbine-powered aircraft.

BELOW LEFT Typical of a military powerplant installation is that of the F-14A Tomcat fighter, with a pair of Pratt & Whitney TF30–P–412 afterburning turbofans in the rear fuselage.

ABOVE RIGHT The small size and fuel economy of the RB199 afterburning turbofan is a primary reason for the compact overall dimensions of the Panavia Tornado.

RIGHT Readily apparent in this illustration of a Tornado GR Mk 1 in flight are the buckets of the nozzle-mounted thrust-reversers in the stowed position.

ABOVE A unique STOVL powerplant arrangement is that of the Yakovlev Yak-38A, with a vectored-thrust Lyulka AL-21F turbojet in the aft fuselage and two directlift Koliesov ZM turbojets in the forward fuselage.

BELOW Yak-38 combat aircraft on the flightdeck of the Soviet carrier *Minsk*.

STOVL NOZZLES

The Harrier's unique four-poster nozzle arrangement taps high-pressure cool air from the compressor for ejection through the forward pair of vectoring nozzles while hot gas from the combustion stage is ejected through the aft pair of vectoring nozzles. The Pegasus turbofan has proved remarkably successful in this role, but is now nearing the limits of its design potential in its current form.

YAKOVLEV YAK-38 FORGER-A

TYPE: single-seat carrier-borne fighter and multi-role combat aircraft with STOVL (short take-off and vertical landing) capability
WEIGHTS: empty 16,281 lb/7,385 kg; maximum take-off 25,794 lb/11,700 kg for vertical take-off or 28,660 lb/13,000 kg for short take-off
DIMENSIONS: span 24 ft 0¼ in/7.32 m; length 50 ft 10⅓ in/15.50 m; height 14 ft 4 in/4.37 m; wing area 199.14 sq ft/18.50 m²
POWERPLANT: one 17,989-lb/8,160-kg thrust Lyul'ka AL-21F non-afterburning turbojet and two 7,870-lb/3,750-kg thrust Koliesov ZM non-afterburning turbojets, the former fitted with two vectoring nozzles and the latter being designed to operate purely as lift jets
PERFORMANCE: speed 627 mph/1,110 km/h; ceiling 39,370 ft/12,000 m; range 460 miles/740 km
ARMAMENT: up to 7,937 lb/3,600 kg of disposable stores (including AA-2 Atoll and AA-8 Aphid short-range air-to-air missiles, rocket pods, cannon pods, air-to-surface missiles and free-fall bombs) on four underwing hardpoints

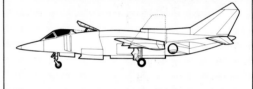

In the shorter term greater power can be derived from the introduction of plenum-chamber burning, a form of afterburning in the cool airflow before it emerges from the front nozzles. Greater long-term benefit may accrue from the use of a three-poster propulsion arrangement, which would allow plenum-chamber burning on the two lateral nozzles fed from the compressor, and afterburning on a single aft nozzle fed from the combustion stage. The tactical advantages of such a system would be considerable: greater thrust could be generated and drag would be reduced, while the afterburning would be maximized in efficiency by a straight-through jetpipe without the current pair of right-angle bends.

Although thrust-vectoring is currently used only on STOVL aircraft, it has great potential on more conventional aircraft. The exhaust of current conventional aircraft is directed through circular nozzles along or close to the aircraft's longitudinal axis to drive the aircraft forward. But in recent years designers have explored the possibility of using a two-dimensional nozzle arrangement to vector this thrust and so produce greater agility in combat aircraft. A new version of the F-15 is being prepared with rectangular nozzles that can be used to vector the thrust up or down: this will help to reduce the take-off and landing runs to a marked degree (in collaboration with canard foreplanes), but will also generate greater aerial agility by introducing a turning moment to the axial force currently provided by the exhaust.

The further possibility exists of vectoring the thrust laterally as well as vertically on widely separated engines such as those of the F-14, with differentially vectoring nozzles to aid control in roll. There is no way that thrust-vectoring could effectively replace aerodynamic control surfaces, but there is every reason to press ahead with the development of such a system to supplement the conventional flying controls of combat aircraft.

INLET EFFICIENCY

At the other end of the engine, the inlet system is the feature that decides the efficiency of the engine installation and therefore that of the aircraft's overall performance. Up to transonic speeds the inlet need not be complex, but thereafter its design becomes increasingly critical as speeds approach and pass Mach 2. The difficulty stems from the designer's natural desire to optimize the airflow to the engine in terms of mass, speed and configuration under all flight regimes. Whereas early jet fighters had simple inlets – plain circular or oval types on the North American F-86 Sabre and the MiG-15, or twin triangular units in the wing roots of the Hawker Hunter – supersonic aircraft featured inlets with fully variable areas and configurations, not to mention spill doors and auxiliary inlets, such as the lateral units on the F-4 Phantom II. Enormous ingenuity went into the design of such inlets, which reached

TOP Thrust-to-weight ratio is particularly important in the engines of VTOL aircraft, and considerable effort is being devoted to an upgraded Pegasus turbofan with plenum-chamber burning to boost thrust.

ABOVE The Pegasus is characterized by short length but considerable diameter.

their apogee on high-performance fighters such as the F-15, but it was recognized all along that they were both costly and complex, ultimately increasing aircraft weight and functional complexity.

Should the real utility of the inlet be doubted, it is sufficient merely to remember that on engines such as the SR-71's Pratt & Whitney JT11Ds and the MiG-25's Tumanskii R-31s some 70% of the thrust at maximum speed is provided by the inlets. Given the efficiency of modern inlets and nozzles, it is arguable that the ideal would be to replace the intervening turbine with an efficient ramjet to convert the inlet's offering into the gas needed by the nozzle. This is a nice idea, though it begs the question of how to power the aeroplane at speeds below the ramjet's effective operating speed (typically Mach 3).

The far-reaching analysis of combat aircraft undertaken by the US Air Force in the late 1960s and early 1970s to assess the implications of air warfare over Vietnam revealed how little time a Mach 2 fighter spent at Mach 2. The proportion may be only 0.1%, and while it may help the fighter to close on its prey before an engagement, Mach 2 capability is thereafter useless. Combat almost immediately degrades

LEFT A typical early-generation engine installation is evident on this Commonwealth Aircraft Sabre; the Rolls-Royce Avon non-afterburning turbojet is aspirated via a plain nose inlet and exhausts via a plain nozzle.

BELOW LEFT Altogether more complex is the arrangement of the F-15's two Pratt & Whitney F100 afterburning turbofans.

RIGHT An F-15 climbs in full afterburner.

BELOW The fuselage shell of the IAI Lavi multi-role fighter reveals the complexity of modern design and construction even before the addition of the engine and avionics.

BOTTOM The Lavi, after the installation of its Pratt & Whitney PW1120 turbofan.

performance to speeds of around Mach 0.8, at which the fixed inlet is more than adequate if properly designed.

Inevitably, therefore, the fully variable inlet has lost much of its attraction, and its elimination has simplified the task of the designer, manufacturer and maintenance man, as well as easing the burden on the taxpayer. Modern combat aircraft have plain inlets, often under the forward fuselage for optimum efficiency at high angles of attack. This means that speeds in excess of Mach 2 are no longer readily attainable, but the arrangement is more than satisfactory at speeds below Mach 2.

STEALTH TECHNOLOGY

Another feature of variable-geometry inlets that is a liability in the combat arena is their great radar cross section (RCS), the geometry of the inlets and their airflow-controllers involving an angularity that reflects radar energy beautifully. There is currently much emphasis on reducing the radar, electronic, thermal and acoustic signatures of combat aircraft. The whole field of stealth technology, as this science is known, is highly classified, and even the existence of a supposed stealth aircraft – the Lockheed F-19 reconnaissance fighter – is denied by the American authorities.

It is undeniable, however, that considerable strides have been made in the development of RAM (radar-absorbent material) technology

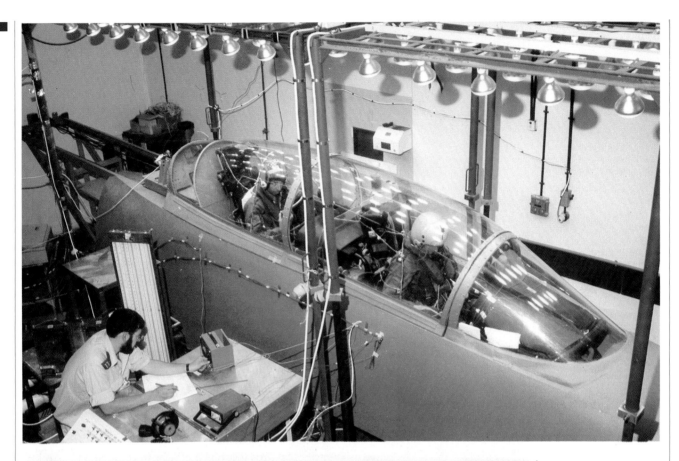

ABOVE Something of the electronic and systems complexity of modern combat aircraft is conveyed by this test rig, designed to evaluate the Lavi's environmental control and cabin pressurization systems.

LEFT, ABOVE RIGHT AND BELOW RIGHT The Boeing B-52G Stratofortress belongs to an earlier generation of combat aircraft, the modest power of contemporary turbojets requiring the use of eight such Pratt & Whitney J557-P-43 engines in four pairs under the wings, where their weight helped to reduce the bending on the wings. The powerplant system has remained essentially unaltered, but the B-52G's electronics have evolved through at least six phases to keep them as modern as possible, and evidence of this trend is seen in the forest of exterior antennae and fairings over the fuselage. The two blisters under the nose accommodate the low-light-level and forward-looking infra-red sensors of the variant's Electro-optical Viewing System for low-level penetration of enemy airspace.

low-RCS airframes. RAM is another highly classified subject, but the basic concept is comparatively straightforward: such materials are used in the structure or as coatings over the structure to absorb electromagnetic radiation. They enable the potential target aircraft to soak up the electromagnetic radiation broadcast by hostile radar systems, reflecting so little that the echo is difficult for the enemy to detect and then plot.

Inevitably, there are areas where the RAM is unsuitable. Here it is the designer's responsibility to reduce the reflective nature of the surface, or to build electromagnetic traps into it. A good contrast in such technology is provided by the US Air Force's heavy bombers, the Boeing B-52 Superfortress and Rockwell B-1B. The B-52 was designed in the late 1940s and early 1950s without any thought of reducing its RCS and has a massive slab-sided fuselage to

ABOVE LEFT The Rockwell B-1A was designed for supersonic penetration of enemy airspace at high altitude, and as such its powerplant installation of four General Electric F101-GE-100 afterburning turbofans was centred on variable geometry for optimized airflow conditions at all altitudes.

LEFT The production version of the B-1 is the B-1B low-level subsonic penetration bomber, and its F101-GE-102 afterburning turbofans are part of a simpler powerplant arrangement with fixed-geometry inlets but variable-geometry nozzles.

ABOVE Mock-up of the B-1's Offensive Avionics System operator station, with screens for the electro-optical viewing system (above) and radar (below), the latter controlled by the operator's joystick. In front of the operator is the alphanumeric screen and keyboard for interface with the system's computers, and to his left are the displays and controls for stores management and navigation.

which the wings, tailplane and rudder are butted at the maximum radar-reflecting angle of 90°. To make matters worse its engines are located in short nacelles that allow radars to see the compressor face of each engine without difficulty.

The B-1A was designed as the B-52's supersonic high-altitude successor, and achieved an RCS one-tenth that of the B-52 through the incorporation of blended contours and curved surfaces that reflected incoming radiation in a number of directions and so reduced that directed straight back at the emitter. When the B-1A metamorphosed into the B-1B, a low-level penetration bomber with lower speed, the original variable inlets were replaced by plain inlets with streamwise baffles to reduce the B-1B's RCS to a mere one-tenth of that of the B-1A – just one-hundredth of that of the B-52.

The design of the B-1B's inlets is a model for future development, the internal arrangement of the inlet and the baffle ensuring that radar cannot see the compressor face, the electromagnetic radiation that enters the inlet being bounced about inside it and dissipated rather than reflected back to the emitter. The same basic approach has been taken with stealth aircraft such as the SR-71 strategic reconnaissance aircraft, whose very structure and design were conceived to reduce radar reflectivity: much of the SR-71's exterior surface is made of a RAM honeycomb, while its internal structure is arranged without right angles wherever possible, the use of small angles trapping the electromagnetic radiation into a series of energy-sapping reflections inside the structure.

The new generation of combat aircraft feature an increasingly curvaceous outer appearance. The tendency is bound to increase as the advantages and techniques of stealth become more evident, and must soon involve the relocation of inlets in less obtrusive spots – perhaps as flush dorsal installations – as well as the adoption of internal weapon bays. At the same time, careful consideration of nozzles may help to reduce the acoustic and thermal signatures of modern combat aircraft, while the reduction of maximum speeds is also playing its part in reducing thermal signature by minimizing aerodymamic heating.

3
PROTOTYPES

AMERICAN BOMBERS

The bomber emerged from World War II as the world's most important strategic weapon. The two events which made that fact abundantly clear were the attacks on the Japanese cities of Hiroshima and Nagasaki, each destroyed in August 1945 by a single Boeing B-29 Superfortress delivering one atomic bomb on each target. The combination of long-range bomber and atomic weapon was clearly established as the arbiter of military power on a global basis, and the USA embarked on an ambitious programme of strategic bomber development.

The initial steps had already been taken, for the USA had realized even before its entry into World War II that although the B-29 would be far superior to its predecessor, the Boeing B-17 Flying Fortress, it would be limited in range when carrying its maximum 20,000-lb (9072-kg) bombload. In April 1941 the US Army Air Forces issued a specification for a bomber able to carry a maximum bombload of 72,000 lb (32,659 kg), or in more realistic terms a 10,000-lb (4536-kg) bombload over a radius of 5000 miles (8047 km). Speed was to be between 300 and 400 mph (483 and 644 km/h), and a service ceiling of 35,000 ft (10,670 m) was to be attained.

A number of designs were submitted to this requirement and the winner was selected in November 1941, the month before the USA's entry into World War II. It was the Consolidated Model 37, which was ordered in prototype form as the XB-36 finally built by Convair, as Consolidated became known after its merger with Vultee. The design was

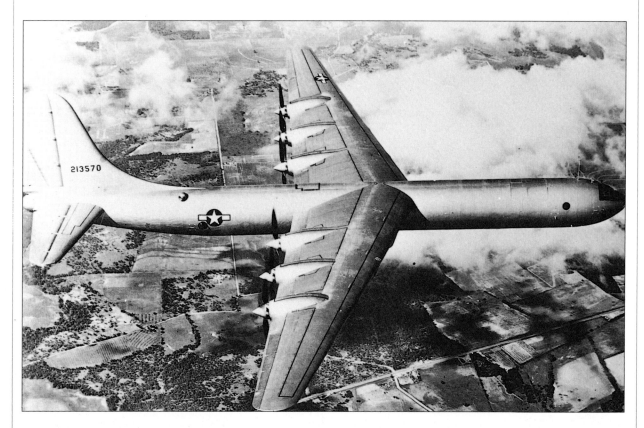

ABOVE. Seen here in its XB-36 prototype form, the Convair B-36 marked the high point in the 'conventional' strategic bomber concept, with an outsize airframe and six powerful radial engines buried in the thick wings to drive pusher propellers.

basically conventional by the standards of the time, but with a span of 230 ft (71.10 m) the aeroplane was exceptionally large. The type did introduce some novel features, however, including a slightly swept wing, a fuselage whose two main pressurized compartments were connected by an 80-ft (24.4-m) tunnel containing a wheeled cart, and propulsion by six pusher propellers driven by radial engines buried in the thick wings.

The prototype programme was initialy slowed by World War II's demands for current aircraft, but then placed at the highest priority in 1943 when the USA realized that strategic blows against Japan could only be struck by long-range strategic bombers. The XB-36 was too late for World War II, and first flew in August 1946. The type was then built in some numbers for the newly formed Strategic Air Command. The B-36 production versions were prodigious aircraft in terms of payload and range, but became the mainstay of the Strategic Air Command's long-range heavy strategic force at the time when the USSR was beginning to field turbojet-powered fighters of high performance armed with heavy cannon and, slightly later, air-to-air missiles. Great effort was therefore put into development of the B-36 with features such as four turbojet engines in podded pairs under the wings to boost performance in speed and altitude.

The last B-36J was delivered in September 1953 and the ultimate operational B-36 was retired in August 1959. The B-36 is therefore the type against which other nuclear-armed bombers should be judged.

The Americans appreciated from the beginning of the B-36's flight-test programme that this mighty type generally met its brief, but that the requirement initially set was obsolescent by the time the XB-36 first flew. The advent of atomic weapons and jet propulsion had altered the concept of

BELOW. Conceptually more advanced than the B-36, the Northrop XB-35 was a low-drag flying wing with four piston engines in the wings driving contra-rotating pusher propeller units.

strategic air power radically, and the B-36 became SAC's mainstay not through any real merits of the design but for lack of anything better until an ambitious development programme had started to yield results.

The main competitor to the B-36 was an ambitious flying wing from Northrop Aircraft Inc. Jack Northrop was a firm believer in the superiority of the flying wing over the conventional aeroplane, whose fuselage and tail surfaces are so weighty and produce so much drag that performance is inevitably degraded. During the late 1920s and the 1930s Northrop had produced a number of experimental flying wings, and in 1941 developed a flying wing design to compete with the Consolidated Model 37. The type was ordered in prototype form as the XB-35. This could carry a maximum bombload of 56,000 lb (25,402 kg), or a 20,000-lb (9072-kg) bombload over a radius of only 2500 miles (4023 km), but with only four engines driving contra-rotating propellers behind the trailing edges of the 172-ft (52.43-m) span wing, was faster than the XB-36 (especially at lower altitude), possessed a usefully higher service ceiling, and was also considerably more agile than the XB-36.

The XB-35 first flew in June 1946, but because there was considerable resistance to the innovative flying-wing design the production contract was later cancelled. There had been problems with the prototype, but these were concerned with factors such as the propeller control mechanisms rather than the structure and flight characteristics of the basic airframe. With the B-36 slated for production, it was then decided to use the design for the evaluation of jet power in a strategic bomber.

The second and third YB-35 pre-production prototypes were therefore converted into YB-35B aircraft: their four 3250-hp (2423-kW) Pratt & Whitney R-4360 piston engines were replaced by eight 4000-lb (1814-kg) thrust Allison J35-A-5 turbojets, a quartet of the jets being

grouped in each trailing edge and aspirated through the same arrangement of leading-edge inlets used to supply carburation and cooling air for the piston engines of the original aircraft.

The YB-35B was redesignated YB-49 while it was being rebuilt, and the first aeroplane flew in October 1947. Speed was increased dramaticaly from 393 to 520 mph (632 to 837 km/h), but such was the thirst of the turbojets that range was halved. There were also several control problems that made the type unsuitable for use as a free-fall bomber, and it was decided to transform the type into a strategic reconnaissance aeroplane. The

Despite its swept flying surfaces and pylon-mounted turbojet engines in side-by-side pods, the Convair YB-60 was clearly a derivative of the obsolescent B-36 and proved inferior to the Boeing XB-52 in both basic performance and anticipated 'developability'.

programme finally succumbed to a combination of technical and political pressures, and was cancelled in April 1949. This was perhaps a sound move for the wrong reasons, for in June 1949 and March 1950 the first and second YB-49s were lost in mid-air explosions.

The YB-49 story seemed to show that while they offered the possibility of much improved performance, jet engines were too thirsty for use in heavy strategic bombers and seemed to affect handling characteristics. Even so, the US Air Force remained convinced that the performance offered by the jet engine must eventually be matched with the payload and range offered by the piston engine to produce a new generation of heavy strategic bomber. In April 1945 the USAF's predecessor, the USAAF, had issued its specification for a B-35/B-36 replacement with turbine propulsion, and the USAF pushed this programme with considerable vigour.

The solution to the payload/range equation was already suspected in the USA during World War II, but it was in German research data captured at the end of that war that the proof was found. High speed and great range could be provided by the jet engine in combination with swept flying surfaces. Thus both the major submissions to the April 1945 requirement featured swept flying surfaces and, by unusual coincidence, power provided by eight turbojets in four twin-jet packages in nacelles cantilevered below and forward of the wings. Such pod-mounted installations offered considerable structural advantages in evening loads out along the wing, reduced the tendency of the wing to twist and, as the primary design consideration, left the engines in an easily accessible location. Additionally, such an installation made it simle to adapt the pod for different and more efficient engine types as these became available.

The failing submission to the 1945 requirement was the Convair YB-36G, which was in essence the B-36 recast with flying surfaces swept at 35° and with a powerplant of eight Pratt & Whitney J57-P-3 turbojets. The type was redesignated YB-60 before two prototypes were ordered in March 1951, and the first of these flew in April 1952. The type offered a 75 per cent structural commonality with the B-36, but in addition to its new flying surfaces had a revised landing gear arrangement and considerably larger fuel capacity. Maximum speed was 550 mph (885 km/h) at 55,000 ft (16,763 m), but in all operational aspects the competing Boeing aeroplane was superior, as well as offering far greater development potential as it was an all-new design.

The XB-52 was designed by Boeing, and exemplifies the rapid evolution in the second half of the 1950s of thoughts into flight at high speed over long ranges with a substantial payload. In July 1948 Boeing received a contract for two examples of its initial design, a large machine with wings swept at 20° and powered by turboprop engines that seemed to offer the right compromise between high speed and long range. By the end of the year, however, Boeing had been able to absorb the implications of some captured German research data, and revised the design as the Model 464 with wings swept at 35° and a powerplant of eight turbojets. To this extent the XB-52 became the 'big brother' of the existing B-47 Stratojet, a medium strategic bomber of quite exceptional aerodynamic cleanliness and flight capabilities. Another feature adopted from the B-47 was the unusual landing gear, an arrangement of main units under the fuselage (two sets of side-by-side units in a 'bicycle' disposition) with balancing outrigger units under the wings.

The B-52 proved to be an exceptional heavy strategic bomber, and production was undertaken in variants up to the B-52H with steadily increased power and electronics for the carriage of heavier warloads through an increasingly effective defence. The B-52 series remained the pre-eminent manned bomber of SAC until the late 1980s, when it was overtaken by the Rockwell B-1B. Even so, B-52G and B-52H models remain in valuable service as carriers for air-launched cruise missiles.

Though a long-term protagonist of the heavy bomber for strategic-level use, the USAAF was also a firm advocate of the medium bomber for tactical- and operational-level use. The Martin B-26 Marauder and North American B-25 Mitchell served magnificently in these roles right through World War II, but by 1944 the USAAF had become confident of the need to develop turbojet-powered successors. The 1944 requirement demanded a speed of at least 500 mph (805 km/h), a tactical radius of 1000 miles (1609 km) and a service ceiling of 40,000 ft (12,190 m). By December of the same year proposals had been received from Boeing, Convair, Martin and North American.

ABOVE AND BELOW. The keynotes of the Consolidated XB-46 were extremely refined aerodynamic lines and turbojet propulsion in an airframe of basically conventional concept. Noteworthy features are the fighter-type cockpit canopy and extremely fine fuselage lines.

The Boeing design was eventually revised into a swept-wing type with underslung podded engines, and in this form matured as the remarkable B-47 Stratojet medium strategic bomber, the type that served alongside the B-36 until both were supplanted by the B-52.

Even though its capabilities removed the B-47 from the tactical/operational level to the strategic arena, the USAAF was still interested in the more limited type, and the other three 1944 submissions reached hardware form. The Convair design materialized as the XB-46 that first flew in April 1947.

The type was exceptionally clean in aerodynamic terms, and was powered by four 4000-lb (1814-kg) thrust Allison-built General Electric J35-A-3 turbojets located as pairs in each wing. The Martin offering was the XB-48, which first flew in June 1947. Like the XB-46, this was a straight-winged type and, powered by six Allison-built J35-A-5s located as triplets in each wing, failed to achieve the performance required. The third submission was the XB-45 from North American. This had been regarded from the beginning as an interim type of less advanced aerodynamic concept than the other two submissions, but proved effective and was ordered into limited production as the B-45 Tornado.

ABOVE. The Douglas XB-43 was an ambitious design intended to produce a tactical bomber of excellent performance without undue technical risk, and notable features are the compact fueselage accommodating the retractable tricycle landing gear, and the flush lateral inlets for the turbojet powerplant.

In World War II the USAAF also made widespread and effective use of the attack bomber for battlefield tasks, and the most striking examples of the genre were a pair of Douglas aircraft, the A-20 Havoc and its successor, the A-26 Invader. For post-war service the USAAF wanted a higher-speed successor, and the specification produced two fascinating types in the Douglas XB-43 and Martin XB-51. The XB-43 was a development of the wartime XB-42 Mixmaster. This was based on two 1800-hp (1342-kw) Allison V-1710-25 inline engines located in tandem in the fuselage to drive a pair of contra-rotating propellers behind the cruciform tail unit. The second aeroplane was completed in XB-42A form with a pair of underwing turbojets to boost performance. The concept was taken a step further in the XB-43, which first flew in May 1946 with a pair of General Electric J35-GE-3 turbojets aspirated via flush lateral inlets just below and behind the cockpit. The flight test revealed good performance but a number of handling limitations, so the type was not ordered into production.

The XB-51 was an altogether more ambitious type. It featured a tandem-unit landing gear arrangement, a thin variable-incidence wing whose leading edges were swept at 35°, a swept T-tail, and a powerplant comprising three General Electric J47-GE-13 engines located as one in the tail and the other two in individual nacelles under the forward fuselage. The

first of two aircraft flew in October 1949, and although flight trials confirmed that the XB-51 had very good performance, they revealed that the type had poor handling qualities in the air. Like the XB-43, therefore, the XB-51 did not proceed past the prototype stage and the USAF opted instead for licence-production of the English Electric Canberra as the Martin B-57.

In the 1950s, the technical solution to Soviet air-defence capabilities seemed to the Americans to lie with bombers that could fly both higher and faster. This, the USAF reasoned, would make it impossible for effective anti-aircraft fire to be brought to bear, and would reduce to a minimum the chances that a manned fighter could be vectored into an intercept position. Given its belief in the technical solution, the USAF opted for the development of supersonic bombers. The role of the B-47 was eventually taken by the Convair B-58 Hustler, an extraordinary tail-less delta with four afterburning turbojets mounted in pods below the wings. To keep fuselage cross section to a minimum and boost performance on the given power, the type was designed for employment with a large ventral pod containing the fuel for the outward leg of the mission, together with the nuclear warload. It was jettisoned over the target.

A supersonic, indeed Mach 3, successor was also proposed for the B-52. This was developed as the North American XB-70 Valkyrie, which first flew in prototype form during September 1964.

The airframe made extensive use of contemporary 'exotic' alloys to overcome the problems associated with kinetic heating. The wings were swept back at 65° 34' on the leading edge, and were covered with brazed stainless steel honeycomb panels welded together to produce an extremely strong yet heat-

BELOW. The Douglas XB-42 was predecessor to the XB-43, and secured very good piston-engined flight performance through a combination of clean design and a pusher propeller. This XB-42A development added a pair of small turbojets under the wings for boosted performance.

LEFT. The Martin XB-51 was a particularly ambitious design with all-swept flying surfaces, including a T-tail, and three turbojets in a most unusual configuration of two under the forward fuselage and the third buried in the tail and aspirated via an inlet at the forward edge of the fin's long dorsal extension.

resistant whole. Similar construction was used for the huge rectangular engine duct under the centreline, the twin vertical tail surfaces and part of the fuselage. The advanced aerodynamics of this elegant yet menacing warplane were based on a large delta wing from whose centre grew a slim forward fuselage complete with canard foreplanes. The powerplant comprised six 31,000-lb (14,062-kg) thrust General Electric YJ93-GE-3 afterburning turbojets in a ducted arrangement under virtually the full chord of the delta wing. Its outer portions were arranged to hinge downward in flight under hydraulic power to improve stability and manoeuvrability. An anhedral angle of 25° was used for low-altitude supersonic flight, increasing to 65° for high-altitude flight at Mach 3. Control was provided by a combination of flaps on the canard foreplanes, no fewer than 12 wide-chord elevons across much of the trailing edge of the wings outboad of the variable-geometry engine exhausts, and large rudders on each of the vertical surfaces. Control of so complex an aerodynamic platform moving at high supersonic speeds was effected with the aid of a three-axis stability-augmentation system.

Even before the fist prototype flew, however, technological developments in air defence had made the XB-70 obsolete.

BELOW. Contrasted with the B-36 behind it, the Boeing XB-52 prototype of the Stratofortress strategic bomber reveals the advanced aerodynamic thinking that made this so great a warplane. The heavily framed fighter-type cockpit enclosure was abandoned in the production model for more conventional flightdeck glazing.

■ The destruction of a high-flying Lockheed U-2 spyplane by a Soviet surface-to-air missile confirmed that the combination of radar, computer and guided missile had made all but suicidal the high-altitude penetration of enemy airspace, even at very high speeds. The two SB-70s were therefore completed and flown as research platforms towards the USA's planned supersonic airliner.

For its time the XB-70 was a stupendous technical achievement, and confirmed the position of North American (soon to become part of Rockwell) as the Western world's foremost designer of very high-speed aircraft. Another machine from the same stable in much the same period was the X-15 hypersonic research aeroplane, a rocket-powered type air-launched by an adapted B-52 bomber within a programme that yielded important results in flight at very high speed and extreme altitudes. The X-15 was built largely of titanium and stainless steel, covered mostly with a so-called 'armour skin' of Inconel X nickel steel alloy to withstand temperatures ranging from +1200°F to −300°F. It should be noted, however, that far higher temperatures were recorded by the X-15A-2 after this type had been fitted with Emerson Electric T-500 ablative material to provide a capability for comparatively steep (and therefore high-friction) angles of re-entry after apogee in the interface between the troposphere and space. This capability to operate on the edges of space demanded a reaction control system for orientation of the aeroplane in these virtually airless regions: a rocket system was used for this stem, with four nozzles in the wingtips and eight more nozzles in the nose to provide three-dimensional manoeuvre capability.

In its later forms the X-15 reached a maximum speed of Mach 6.72 (4532/mph/7297 km/h) and comfortably exceeded its design requirement of a 264,000-ft (80,500 m) ceiling by reaching 354,200 ft (107,960 m), or 67.08 miles (108 km).

ABOVE. Seen on its first flight, the General Dynamics FB-111A gave the US Air Force's Strategic Air Command a potent medium-range strike capability with a supersonic aeroplane.

LEFT. This series of photographs graphically reveals the wings of the General Dynamics F-111 interdictor in their minimum, intermediate and maximum sweep positions.

Given its experience with the XB-70 and X-15 programmes, it is not surprising that Rockwell was selected as prime contractor for the aerospaceplane glider component of the US space shuttle system.

In terms of military aircraft, however, such capability in the realms of hypersonic high-altitude flight were irrelevant and indeed dangerous, given the growing sophistication of the Soviet air defences likely to be encountered in any major confrontation between the superpowers. The USA was nonetheless determined to maintain a manned bomber force as one of the three 'legs' of its strategic nuclear triad, the other limbs being the intercontinental ballistic missile force and the submarine-launched ballistic missile force.

As already noted, the solution in the short term was the modification of the B-52 for low-altitude penetration of Soviet airspace at high subsonic speed while a major programme was launched to create a supersonic penetration bomber. This was the Rockwell B-1, which resulted from a November 1969 requirement for a medium-altitude with dash capability of Mach 2.2+ for the high-speed delivery of free-fall and stand-off nuclear weapons. Submissions were received from several companies, Rockwell's design being selected in 1970 as the B-1A. The full-scale design and development programme for the initial production version was soon under way. The initial model was a complex and highly advanced variable-geometry type with General Electric F101 turbofans and fully variable inlets for maximum capability in all elements of the flight envelope.

The prototype first flew in December 1974, and the flight test programme moved ahead. In June 1977, however, President Carter decided to scrap the programme in favour of cruise missile development, although flight trials with the B-1A aircraft were to be continued for research purposes. Then with the inauguration of President Reagan matters began to look up again, the new

ABOVE. An artist's impression provides a nice insight into one of the forms the proposed FB-111H strategic bomber could have taken with a much refined fuselage/powerplant combination in a more refined airframe.

LEFT. The Rockwell B-1A was planned as a supersonic successor to the B-52 with variable-geometry wings and advanced electronic systems, and reached the stage of prototype trials with remarkably few problems. The project then foundered on political objections.

administration deciding during October 1981 to procure 100 examples of a much revised B-1B version in the low-level penetration role for high-subsonic delivery of free-fall and stand-off weapons. The B-1B was therefore a straightforward but nonetheless major adaptation of the B-1A optimized for the low-level transonic role with fixed inlets and revised nacelles (reducing maximum speed to Mach 1.25). But it did have a strengthened airframe and landing gear for operation at higher weights with nuclear and conventional weapons over very long ranges. Other changes were concerned with reduction of the type's already low radar signature, S-shaped ducts with streamwise baffles being adopted to shield the face of the engine compressors, and radar absorbent materials being installed in sensitive areas to reduce electromagnetic reflectivity. The second and fourth B-1As were used from March 1983 to flight-test features of the B-1B, which first flew in September 1984 with the advanced offensive and defensive electronic systems.

From the beginning the B-1B was seen as an interim but valuable operational asset bridging the gap between the obsolescent B-52 Stratofortress and a new strategic bomber. As has been noted, one of the major advantages of the B-1B over the B-52 is its low 'observability' in the electromagnetic and thermal senses, this being estimated as being perhaps only one-tenth of the B-52's 'observability'. This process of increasing 'stealthiness' was to be taken considerably further with the B-1B's successor, which possessed only one-tenth of the B-1B's 'observability' to produce a figure a mere one-hundredth of the B-52's electromagnetic and thermal signatures.

The resulting aeroplane is the Northrop B-2, which was developed at enormous cost during the late 1970s and 1980s, and revealed in November 1988 for an initial flight in July 1989. Unlike the low-altitude B-1B, however, the type is designed for medium- and high-altitude penetration of hostile airspace, relying on its stealth design and composite structure to evade detection by enemy air defence systems until it has closed to within a few miles of its target. The aeroplane is a flying wing with 40° swept leading edges and W-shaped trailing edges featuring simple flight-control surfaces: elevons for pitch and roll control, and 'differential drag' surfaces for yaw control. These are operated by a fly-by-wire control system to ensure optimum control responses in this design of relaxed stability intended for positive aerodynamic control at all times. Throughout the airframe, emphasis is placed on completely smooth surfaces (especially in the area of the blended flightdeck and the engine nacelle bulges) wherever hostile radar emissions may hit its surfaces. In combination with a structure largely of radar-absorbent materials, this ensures that an absolute minimum of radiation is reflected back toward the hostile emitter/receiver system.

Radar reflectivity is very low because of the overall design which included a computer-designed interior structure that ensures the entrapment rather than reflection of penetrating radiation and use of radiation-absorbent materials. The shape (including shielded upper-surface inlets) has been carefully optimized by computer design and computer-controlled manufacture to produce a head-on radar cross section of about 10.76 sq ft (1 m^2) by comparison with the 107.64 sq ft (10 m^2) of the B-1B and the 1076.37 sq ft (100 m^2) of the B-52. To supplement this reduction of the B-2's electromagnetic 'observability', the design is optimized for reduction of the thermal signature, which could otherwise be detected at quite considerable range by the infra-red search and tracking systems carried by the latest generation of Soviet fighters such as the Mikoyan-Gurevich MiG-29 and Sukhoi Su-27. The reduction

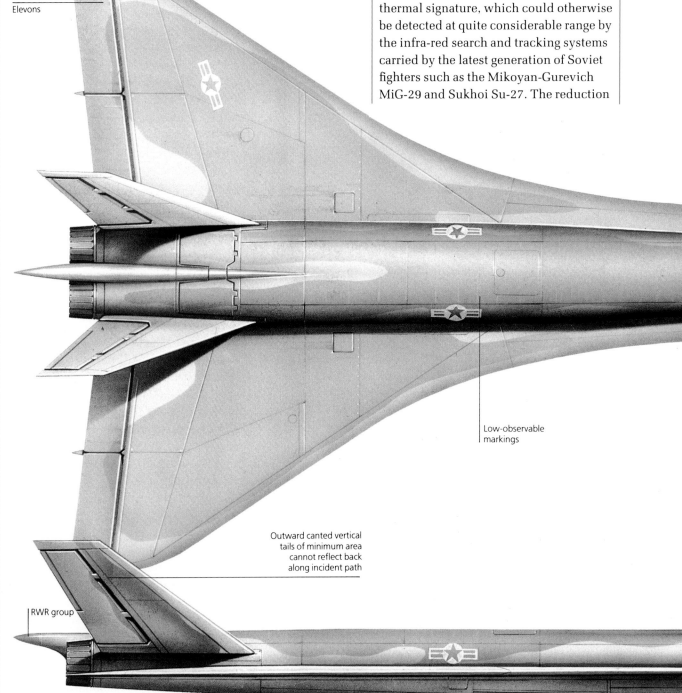

Elevons

Low-observable markings

Outward canted vertical tails of minimum area cannot reflect back along incident path

RWR group

of thermal 'observability' is ensured by the use of non-afterburning turbofan engines and specially conceived two-dimensional outlets where the mixing of hot exhaust gases with cold freestream air before release significantly reduces both the thermal and acoustic signatures.

All this results in an aeroplane which, the Americans hope, will be virtually undetectable by foreseeable radar and infra-red sensors except at the very short ranges at which the type will be seen with the naked eye. Production of 132 B-2s is planned, and these will be the carriers for 2000 of the 4845 strategic nuclear weapons in the US Air Force's inventory. The US Department of Defense's 1989 review of financial commitments in the face of the USA's enormous budget deficit included among its provisional proposals that the service debut of the B-2 be postponed for at least one year to save on financial outlay and to allow additional time for development of this extremely ambitious and complex aeroplane.

It is worth noting here that the USA is devoting considerable efforts to the defeat of 'stealth' technology. Thus, the Department of Defense reasons, the US forces will make headway into the technologies for 'seeing' Soviet aircraft approximating the low 'observability' of current American 'stealth' aircraft, simultaneously paving the way for the creation of more advanced 'stealth' technologies to maintain the USA's current lead over the USSR in this field. A number of advanced concepts are being evaluated, including some as roundabout as a satellite surveillance system to watch for the shadows cast onto the surface of the Earth by 'stealth' aircraft and missiles.

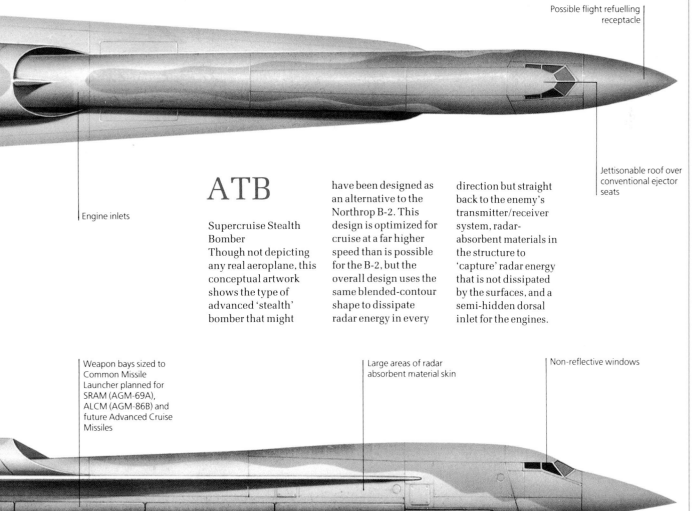

ATB

Supercruise Stealth Bomber
Though not depicting any real aeroplane, this conceptual artwork shows the type of advanced 'stealth' bomber that might have been designed as an alternative to the Northrop B-2. This design is optimized for cruise at a far higher speed than is possible for the B-2, but the overall design uses the same blended-contour shape to dissipate radar energy in every direction but straight back to the enemy's transmitter/receiver system, radar-absorbent materials in the structure to 'capture' radar energy that is not dissipated by the surfaces, and a semi-hidden dorsal inlet for the engines.

NORTHROP B-2

Developed at enormous cost from the late 1970s and during the 1980s in a programme of utmost secrecy and technical sophistication, the B-2 was first revealed in November 1988 for an initial flight in July 1989. This extraordinary flying wing warplane was designed as a successor to the Rockwell B-1B in the penetration bomber role. This task demands deep incursion into enemy airspace before release of its primary nuclear weapons, which can be of the free-fall 'dumb' or guided missile types.

Unlike the B-1B, which is designed for high-subsonic penetration at very low altitude, the B-2 is intended for moderate-speed penetration at medium and high altitudes. In this regime the bomber relies on its 'stealth' design and composite structure to evade detection by enemy air defence systems until it has closed to within a few miles of its target. In overall configuration the B-2 is a flying wing with its leading edges swept at 40°, and the W-shaped trailing edges feature simple flight-control surfaces (elevons for pitch and roll control, and 'differential drag' surfaces for yaw control). The layout is of the relaxed-stability type whose lack of inherent 'flyability' requires that full-time control be exercised by a fly-by-wire system based on high-speed digital computers.

In overall design terms, the emphasis is placed on completely smooth external surfaces with blended flightdeck and nacelle bulges. The B-2's radar reflectivity is very low because the internal structure and external surfaces were designed as a whole, with the aid of a computer to ensure that virtually all received radar energies are dissipated away from the enemy's emitter/receiver system, and that all energies that cannot be thus dissipated are trapped inside the B-2's structure. This process is achieved by the use of radiation-absorbent materials and a carefully optimized shape (including shielded upper-surface inlets for the non-afterburning turbofans. A head-on radar cross section of 10.76 sq ft (1 m^2) has been quoted for the B-2 in comparison with 107.64 sq ft (10 m^2) for the B-1B and 1076.37 sq ft (100 m^2) for the elderly Boeing B-52 Stratofortress. Additionally, the careful mixing of hot exhaust gases with cold freestream air before release through the B-2's 2D nozzles reduces the B-2's thermal and acoustic signatures to a very significant degree in this firmly subsonic design. In keeping with this trend of avoiding features that could betray it, the B-2 has been designed with a minimum of emitters sensors such as radar. There is a low-energy radar for use in the terminal stages of the attack, but otherwise the 'electronically silent' B-2 relies on passive electromagnetic systems, navigation with the aid of advanced inertial platforms, and

However, the US Department of Defense's 1989 review of financial commitments in the face of the USA's enormous budget deficit included amongst its suggestions that the service debut of the B-2 be postponed for at least one year to save on financial outlay. This will also allow additional time for development of this extremely ambitious and complex aeroplane. There is considerable technical and political opposition to the bomber, the former focused on suggestions that the B-2 is not as 'stealthy' as its protagonists claim and the latter on the notion that construction of such a warplane would be provocative at a time of easing world tensions.

probably the use of other sensors such as passive optronic or scarcely detectable laser varieties. There can be little doubt that the B-2 is a stupendous technical achievement in a host of technologies from structures and aerodynamics to systems and materials. Production is planned of 132 B-2s, these being the carriers for 2000 of the 4845 strategic nuclear weapons in the US Air Force's inventory.

SPECIFICATION
Northrop B-2

Type: strategic bomber and missile-carrying aeroplane

Accommodation: flightcrew of two with provision for a third man

Armament: this is carried in two weapon bays located side-by-side in the lower fuselage up to a maximum weight of 80,500 lb (36,515 kg) of disposable stores; each bay is thought to accommodate one eight-round rotary strategic launcher for a total of 16 Boeing AGM-131 SRAM-II missiles or 1.1-megaton B83 thermonuclear free-fall bombs; alternative loads are 20 megaton-range B61 thermonuclear free-fall bombs, or 22 1500-lb (680-kg) free-fall bombs, or a large number of smaller bombs

Electronics and operational equipment: communication and navigation equipment, plus Hughes Aircraft covert strike radar with conformal phased-array antennae in the leading edge, and an integrated passive electronic warfare suite

Powerplant: four 19,000-lb (8618-kg) thrust General Electric F118-GE-100 non-afterburning turbofans

Performance: maximum speed 475 mph (764 km/h) or Mach 0.76 at high altitude; operational radius 3800 miles (6115 km) with eight SRAMs and eight B61s

Weight: maximum take-off 371,000 lb (168,286 kg)

Dimensions: span 172 ft (54.43 m); length 69 ft (21.03 m); height 17 ft (5.18 m); wing area not revealed

SOVIET BOMBERS

The American effort in the development of jet-powered bombers described in the last chapter was at least matched and perhaps exceeded by that of the Soviets. They pioneered the concept of the long-range strategic bomber during the 1930s with a number of impressive aircraft from the Tupolev design bureau, but had let this lead slip during World War II when overriding importance was attached to tactical aviation in support of the Red Army in the field. The Soviets did produce the useful Petlyakov Pe-8 four-engined heavy bomber, but this was built only in small numbers.

The success of Allied heavy bombing against Germany and Japan in the closing stages of the war, especially the dropping of atomic bombs on Hiroshima and Nagasaki, persuaded the Soviets to rethink their air policy in favour of strategic bombing. On three occasions Joseph Stalin formally requested that Boeing B-29 Superfortress strategic bombers should be supplied to the USSR under Lend-lease arrangements, but was refused. Tupolev had already started work on a near-copy of the B-29 using data supplied by espionage, and the Shvetsov engine bureau had started to produce plans of the Wright R-3350 engine and its General Electric turbochargers. In the second half of 1944, however, three B-29s force-landed in Siberia and these became pattern aircraft in an extraordinary reverse-engineering programme that saw the creation of the Tu-4 as a virtual exterior clone of the B-29 but with significant interior modifications.

The Tu-4 was placed in service in 1948, and production amounted to about 400 aircraft. This was seen only as a first step and the Soviets were already hard at work on improved bombers. Myasishchyev had planned a much redesigned version as the DBV-202, whose prototype was not completed, while Tupolev moved more cautiously in a number of smaller evolutionary steps. The first result of this programme was the Tu-80 with integral wing tankage and a number of aerodynamic improvements equivalent to those that had turned the basic B-29 into the superior B-50. It first flew in November 1949 but did not proceed past the prototype stage as the considerably more advanced Tu-85 was imminent.

The Tu-85 was a major enlargement of the basic concept with more powerful engines and nearly double the fuel capacity of the Tu-4 in an effort to create an intercontinental strategic bomber able to match the capabilities of the B-36. Work proceeded with impressive speed, and the first prototype was flown in January 1950, a mere two years after the

RIGHT. The Myasishchyev M-4 was a bold and technically very impressive attempt to provide the USSR with an intercontinental strategic bomber. The bomber had very good performance in all parameters but range, which was much lower than specified because of the thirst of the only available turbojet engine, the Mikulin AM-3.

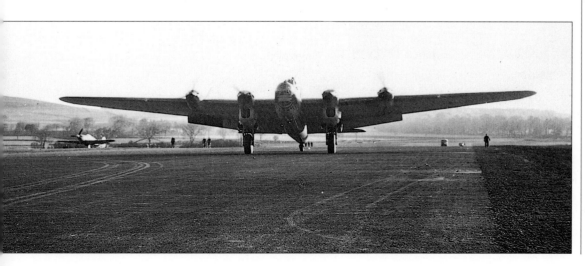

LEFT. The Petyakov Pe-8 entered limited production in World War II as a heavy bomber, and was a conventional aeroplane with notable features such as defensive gun positions in the rear of the inner engine nacelles and, in the prototypes, a fuselage-mounted engine to drive the massive supercharger that supplied a larger volume of air to all four flight engines.

beginning of the programme. Limited flight trials were undertaken, but further development was cancelled in favour of all-jet bombers, a decision that can be questioned in hindsight because it was 1956 before a jet bomber began to enter service and the Tu-85 offered a good combination of performance and payload.

The projects and aircraft evolved by the Tupolev bureau during this period offer a good indication of Soviet aeronautical thinking in the period up to the advent of the USSR's first two turbine-powered strategic bombers, the Myasishchyev M-4 and the Tupolev Tu-95. After the cancellation of the Tu-85 Tupolev realized that the Tu-4 would have to soldier on for several years, and proposed an upgraded version as the Tu-4TVD with either Klimov VK-2 or Kuznetsov NK-4 turboshafts, but this concept failed to find official favour.

The bureau then moved forward to a purpose-designed jet bomber based on the airframe of the Tu-2 piston-engined bomber. This was the Tu-77 with two 5000-lb (2272-kg) thrust Rolls-Royce Nene I turbojets, and first flew in June 1947. The Tu-77 was regarded solely as an interim type, and small numbers were built largely so that bomber crews could gain jet experience. At much the same time the bureau developed the Tu-72 as a contender for the Soviet air force's main light bomber requirement, which was ultimately met by the cheaper and more agile Ilyushin Il-28. In an effort to improve performance the type was revised as the Tu-73, adding a 3500-lb (1588-kg) thrust Rolls-Royce Derwent

BELOW. The Tupolev Tu-80 prototype marked a purely Soviet attempt to produce a strategic bomber modelled on, but superior to, the Boeing B-29 Superfortress, which the Soviets had already copied as the Tu-4.

LEFT. The Tu-77 was a prototype based on the successful Tu-2 high-speed bomber, but designed from scratch with two wing-mounted turbojets as its powerplant. The type first flew in 1947, but no production was undertaken as the design was already obsolescent in a period of very rapid aerodynamic progress.

turbojet in an S-shaped duct in the rear fuselage. Further refinement of the basic concept led to the Tu-79 with a pair of 5952-lb (2700-kg) thrust Klimov VK-1 turbojets for much improved performance. Production of about 500 aircraft followed, most of the operational machines being of the Tu-89 variant used in the maritime attack role with a pair of internally carried torpedoes.

These were all straight-winged aircraft with no pretensions even to high subsonic performance. However, in parallel with these efforts Tupolev used the results of captured German aerodynamic research to produce a higher performance design with wing swept at just over 34°. This was the Tu-82, first flown in February 1949. The type was again powered by the two VK-1 turbojets and attained a maximum speed of 580 mph (934 km/h) at 13,125 ft (4000 m) by comparison with 535 mph (861/km/h) at 16,405 ft (5000 m) attained by the identically engined Tu-79 series. No production of the Tu-82 was seriously contemplated because of the capabilities of the in-service Il-28. Yet it was the bureau's experience with these types that led to a classic design, the Tu-88 that first flew in early 1952 and entered service as the Tu-16 series which is still in widespread use for a number of missile-carrier and reconnaissance roles.

The Tu-16 was produced as counterpart fo the US Air Force's Boeing B-47 Stratojet. To give the USSR a heavyweight punch equivalent to that of the Boeing B-52 Stratofortress the Myasishchyev M-4 and Tupolev Tu-95 were developed, each

BELOW. The Tupolev Tu-95 was developed as a technically less ambitious fallback in case of problems with the Myasishchyev M-4 jet bomber, and featured the unusual combination of swept flying surfaces and propeller propulsion, in this instance by monumentally powerful turboprops derived directly from German research in World War II.

LEFT. When first flown in 1952, the Tu-16 was an extraordinary aerodynamic and structural achievement, and so successful was the Tu-88 prototype that little basic development was needed before the aeroplane entered service. Since that time many variants have been introduced with different weapons and/or electronic systems, but these have left the core airframe/powerplant combination little altered.

with four turbine engines and all-swept flying surfaces. The M-4 was conceived as an exceptionally clean turbojet-powered type with its engines buried in the wing roots, and proved an admirable maritime reconnaissance type even though it lacked the range for its intended intercontinental role. Altogether more successful, and still in production in its upgraded Tu-142 anti-submarine and missile-carrier forms, is the Tu-95. This is an apparent anachronism, for despite its all-swept flying surfaces it is powered by four massive turboprops each driving a huge contra-rotating propeller unit for performance not significantly inferior to that of the B-52.

With these two aircraft types in final development to complement the Il-28 light and Tu-16 medium bombers, the Soviets followed the same tactical logic as the Americans and proceeded along the same path toward the supersonic bomber. The first supersonic bomber to enter service (in about 1961) was yet another design from the Tupolev bureau, the Tu-22. This was produced to fulfil the same role as the Tu-16 but with supersonic dash performance to aid the type in penetrating the increasingly sophisticated air defences possessed by the USSR's putative opponents. More interesting, however, was an effort from the Myasishchyev bureau to create a supersonic strategic heavy bomber to succeed its M-4, but which attempted to push too far beyond the state of the art.

This was the M-50, based on the same aerodynamic tailed-delta formula that had produced the Mikoyan-Gurevich MiG-21 supersonic fighter. The overall design was bold yet basically uninspired.

LEFT. The Ilyushin Il-30 was designed as successor to the important Il-28 turbojet-powered light bomber, and was the first Soviet bomber to exceed 1000 km/h (621 mph). Whereas earlier Soviet aircraft of this type had been powered by engines of British design, or by Soviet engines derived from these engines, the Il-30 had two Lyul'ka turbojets of all-Soviet origin. The unusual landing gear arrangement comprised tandem main units under the fuselage plus outrigger units that retracted into the engine nacelles. No production was undertaken, probably because of the type's long field (take-off and landing) requirements.

It was based on a very long area-ruled fuselage with two four-wheel main landing gear bogies arranged in tandem to retract into the lower fuselage. This was basically the same arrangement as that used in the M-4, as was the use of two twin-wheel stabilizers of the outrigger type that retracted aft into the wingtips. The cropped delta wing was located in the shoulder position with its leading edges swept at 50° inboard decreasing to 41° 30′ outboard. The tail unit was conventional for a supersonic type, with powered all-moving slab tailplane halves and a fin with a powered rudder. The crew of three was accommodated in a pressurized nose compartment on tandem ejector seats behind a V-shaped windscreen whose contours were continued aft of the cockpit by a long dorsal spine stretching as far as the extreme tail. In the first aeroplane the four engines were located on pylons under the wing leading edges, and were probably Koliesov ND-7 or VD-7 units each rated at 28,660-lb (13,000-kg) thrust.

First flight dates between 1957 and 1961 have been quoted, and it is believed that a maximum speed in the order of Mach 1.8 was achieved. By the standards of the day this was a good figure, but the range of 3730 miles (6000 km) without payload was poor. The last of several prototypes, generally known as the M-52, had a different powerplant arrangement: the two inboard engines remained on underwing pylons but were afterburning units each rated at 39,683-lb (18,000-kg) thrust, while the two outboard engines remained non-afterburning units but were arranged on pylons with forward-swept leading edges projecting horizontally from the cropped tips of the delta wings. The M-50/M-52 series failed to progress past the prototype stage.

At the other end of the bomber spectrum there was the light bomber, and here the USSR wished to deploy an Il-28 successor that offered at least transonic and preferably supersonic performance. This need was eventually met by the Yakovlev Yak-28, an evolutionary development from the Yak-25 fighter and capable of dash performance in the order of Mach 1.1.

This was a lightweight type by comparison with two aircraft that failed to secure production orders, the Ilyushin Il-54 and the Tupolev Tu-98. These latter were made possible by the development of the Lyul'ka AL-7 turbojet, an important engine of the axial-flow type with a small diameter and high thrust, the latter achieved only at the cost of high fuel consumption.

The Il-28 had already been upgraded in the Il-30 with the Lyul'ka AL-5 turbojet and 35° swept wings to produce a speed of more than 621 mph (1000 km/h), but this had not been ordered into production. The Ilyushin design bureau then moved to a scaled-up version of the

BELOW. The Ilyushin Il-54 flew in 1955 as the prototype of a supersonic light bomber, and like the Il-30 featured Lyul'ka engines and bicycle main landing gear units allied, in this instance, to swept flying surfaces.

ABOVE. From the Tupolev Tu-102 prototype of 1957 the Soviets hoped to develop a whole series of supersonic warplanes, but the only type to materialize as a production type was the Tu-28P long-range interceptor, which was placed in service as a long-range fighter armed with large air-to-air missiles capable of downing American bombers approaching the USSR by the great circle route over the polar regions.

Il-28 as the Il-46 with an uprated version of the Il-30's powerplant, but this too failed to secure any production commitment. The bureau then moved logically to a fully swept version of the Il-30 that offered transonic capability with the possibility of supersonic speed in a shallow dive.

The Il-54 resulted from a 1953 requirement and first flew in early 1955. The type was typical of Soviet thinking for tactical bombers in the period: an oval-section fuselage with a completely glazed bombardier nose; a fighter-type canopy over the pilot and a barbette controlled by the tail gunner; tandem main landing gear units with stabilizing outriggers, and flying surfaces swept at 55°, these last including high-set wings with the two 14,330-lb (6500-kg) thrust AL-7 turbojets pod-mounted below them. Flight trials confirmed that the Il-54 had transonic performance, with a maximum sea-level speed of 715 mph (150 km/h), but no production was authorized.

The Tu-98 was long believed in the West to be a Yakovlev design, the Yak-42, or even an Ilyushin type. Details are still scarce, but it is believed that the type had swept flying surfaces whose wings were angled at 60° inboard decreasing to 55° outboard, and was powered by two 22,046-lb (10,000-kg) afterburning thrust AL-7F turbojets located laterally on the sides of the area-ruled fuselage. The type clearly drew on the design of the Tu-16 bomber with features of the Tu-28 long-range interceptor, and was capable of Mach 1.17 at high altitude. A first flight was probably made in the initial part of 1956, but no production followed as its advantages over the Tu-16 were relatively small at a time when increasing emphasis was being placed for strategic purposes on the vastly expensive intercontinental ballistic missile programme.

The arena in which the supersonic bomber was thought to offer the best capabilities was that encompassing maritime roles such as reconnaissance and anti-ship attack, the latter with supersonic cruise missiles featuring long range and interchangable warheads (nuclear for area attack on major targets such as carrier battle groups, and HE with radar and/or infra-red homing for pinpoint attack on similar targets). The Tu-22 is capable of Mach 1.5 on the power of the two 30,900-lb (14,000-kg) afterburning thrust turbojets pod-mounted in the angles between the rear fuselage and the vertical tail, but lacks range when carrying a weapon such as the AS-4 'Kitchen' missile.

The solution was an improved derivative with variable-sweep outer wing panels capable of movement between 20° and 55°. The former is used for take-off and landing, much reducing field requirement, intermediate positions for low- and high-speed cruise over longer and shorter ranges, and the latter for maximum dash performance including a speed of Mach 2.3. The device of pivoting the outer wing panels has been used in developments of several Soviet warplanes for minimization of field requirements, and maximization of performance especially in terms of the conflicting fields of range and speed. The type was otherwise modified from the Tu-22 in a number of other features, most notably the installation of two afterburning turbofans in conventional ducts along the fuselage sides, and the first examples of the new type were designated Tu-22M, indicating that it was a modification of the baseline Tu-22. The type was still deficient in range, however, and the definitive Tu-26 model switched to a lower-drag landing gear arrangement.

The typical Tupolev feature of main units retracting aft into large fairing behind the wing trailing edges was replaced by a new system in which the main-wheel bogies retract inward into the fuselage. This was the most obvious sign of a major redesign that left from the original Tu-22M airframe only the inner wing, fuselage barrel and vertical tail structures, and the result is an altogether more formidable warplane.

In technical terms the Tu-26 is still an interim variable-geometry type obviating only some of the worst features of fixed fully swept geometry while conferring some of the advantages offered by a full variable-geometry layout.

The first Soviet warplane with fully variable-geometry wings was the Sukhoi Su-24 long-range interdictor, a type technically and operationally comparable with the General Dynamics F-111 tactical and FB-111 strategic aircraft used by the US Air Force. This in turn paved the way for a larger fully variable-geometry warplane, the Tupolev Tu-160 supersonic

Little has been revealed about the prototype development of the USSR's most capable first-line military aircraft, but considerable research and development effort must have preceded the advent of advanced

types such as this Sukhoi Su-24. Given the NATO reporting name 'Fencer', the Su-24 is the USSR's tactical counterpart to the General Dynamics F-111 and, like that American aeroplane, has variable-geometry wings.

strategic penetration bomber, the USSR's counterpart to the Rockwell B-1B in service with the US Air Force. First revealed during 1981 in the form of US reconnaissance satellite imagery of a prototype at that time designated 'Ram-P', the Tu-160 is larger than the B-1B but has the same type of variable-geometry wing planform with no fixed centre section other than the large-chord wing gloves that allow the wing pivots to be located close to the fuselage. The result, as proved by evaluation and initial service operations, is an unrivalled combination of long-range cruise at up to Mach 0.9 with the wings in the modestly swept position and the supersonic dash performance with the wings in the fully swept position.

Nothing has been revealed about the possible design of a Soviet 'stealth' bomber comparable with the B-2, but it is highly likely that such a type is under development using materials and a flight-control system similar to those of the American aeroplane, but possibly a different layout as the USSR lacks a current exponent of the flying wing design concept.

BOMBERS FROM OTHER COUNTRIES

UK

The only other country to have developed and operated long-range strategic bombers is the UK, which still had a major world role to play in the 1950s. The three types that were fully developed and placed into service were the interim Vickers Valiant and two more advanced aircraft, the delta-winged Avro Vulcan and crescent-winged Handley Page Victor. These were powered by four turbojets, although the two more advanced versions went into B.Mk 2 forms with turbofan power and provision for the Avro Blue Steel stand-off nuclear missile in place of the original models' free-fall nuclear weapons.

Before such costly aircraft were produced, however, their new planforms were evaluated on small-scale research aircraft. The delta wing proposed for the Vulcan was trialed in a number of Avro Type 707 research aircraft, which also tested a number of other features in low- and high-speed variants. The crescent-shaped flying surfaces planned for the Victor were similarly evaluated on another small-scale type, the Handley Page H.P.88.

Even though the Valiant was produced as an interim type, the British felt it essential to develop an alternative in case the Valiant proved a failure in its initial trials. This was the Short SA.4 Sperrin, which conformed to a less demanding specification in terms of speed and altitude over the target. The aeroplane was thoroughly conventional by the structural and aerodynamic concepts of the day, with straight flying surfaces whose wings were set in the shoulder position on a comparatively deep fuselage whose lower portions

BELOW. A contemporary of the Vulcan was the Handley Page Victor, and this bomber's crescent-shaped flying surfaces and T-tail were trialled successfully in scaled-down form on the HP.88 aerodynamic prototype.

accommodated the large nav/attack radar (chin position) and internal bomb bay (central position). One unusual feature was the powerplant, whose four Rolls-Royce Avon turbojets were located in under-and-over pairs on the wings about two-fifths of the way between the fuselage and the wingtips.

The first prototype flew in August 1951, and was soon joined by the second machine. The Valiant then proved itself the better aeroplane, and though no production of the Sperrin was then undertaken, the two prototypes were used for a valuable programme of test flying in a number of roles.

FRANCE

At this time France had no ambitions toward a long-range bombing capability with nuclear weapons, but did wish to build up powerful armed forces to secure a position as arbiter of European affairs. Shortly after the end of World War II France planned its first jet-powered

ABOVE. One of the most interesting and important British warplane programmes was that leading to the great delta-winged Avro Vulcan strategic bomber, and the two Type 698 prototypes are seen here in company with the four Type 707 small-scale aircraft that were used to validate many of the Type 698's features.

bomber as the Aerocentre NC.270 with two Rolls-Royce Nene turbojets to create a warplane that with development could have been comparable to the English Electric Canberra, the Ilyushin IL-28 and the North American B-45 Tornado. The airframe of this light bomber was based on elegantly curved lines with flying surfaces (including a T-tail) of modest sweep. Power was provided by two engines located in bulbed wing roots, and the design envisaged an 11,023-lb (5000-kg) bombload carried internally plus a defensive armament of four 15-mm cannon in a TV-controlled tail barbette. Validation of the design was entrusted to a pair of reduced-scale machines, the engineless NC.271-01 and the rocket-powered NC.271-02. The NC.271-01 flew in 1949, but before the NC.271-02 could be flown later in the same year the parent company folded and the whole NC.270 programme was terminated.

Slightly better fortune attended France's next bomber prototype, the Sud-Ouest SO.4000, also intended for the light bomber role. This was again powered by a pair of Nene turbojets, in this instance located side-by-side in the nicely-streamlined fuselage and exhausting at the tail via long jetpipes; the flying surfaces wre swept at 31° and included a mid-set wing. Production of the full-size aeroplane was preceeded by that of two smaller-scale machines, the SO.M-1 glider and the SO.M-2 powered version with a Rolls-Royce Derwentturbojet. The SO.M-1 first flew in September 1949, but was in fact beaten into the air by the powered model, which first flew in April of the same year. The first SO.4000 flew in March 1951, but achieved only this one flight as the programme was cancelled in the light of the prototype's instability and lack of power. The armament planned for the production version was 4409-lb (2000-kg) of bombs carried internally and, for defence, a remotely controlled barbette at each wingtip carrying a single 20-mm cannon. After the cancellation of the SO.4000, the Sud-Ouest design team turned to a more refined type, the SO.4050 Vautour that eventually entered service in three versions optimized for all-weather interception, close support, and bombing from medium and high altitudes.

ABOVE. The aerodynamic proposed Aerocentre NC 270 bomber was tested in scaled down form as the NC 271.01 glider that was air-launched from a piggyback position above a Languedoc transport aeroplane.

TOP RIGHT. The Avro Type 707C was the two-seat version of this scaled-down Avro Vulcan bomber, and was the prototype used for development of power control systems and electronics.

BOTTOM RIGHT. France's first turbojet-engined bomber was the Sud-Aviation SO.4000 that first flew in March 1951. At this time the French aero industry was still emerging from the doldrums of World War II and, although great strides had been made, the obsolescent design of the SO.4000 reveals that the USA and UK were still far ahead.

AMERICAN FIGHTERS

If it is worth operating a bomber or any other type of offensive warplane, then it is also worth possessing a fighter for the protection of one's own air space against attacks by the bombers and other warplanes flown by any real or potential opponent. This has been the rationale for fighter design since 1915, and with the prospect of nuclear-armed bombers becoming all too apparent in the second half of the 1940s it assumed still greater importance for all countries with aspirations to real military strength. In this instance the threat emerged at much the same time as new propulsion and aerodynamic technologies, resulting in a crop of fascinating fighter prototypes from which emerged a few excellent production types.

Both the USA and the USSR had begun to design turbojet-powered fighters in World War II, but then almost paused after the introduction of straight-winged first-generation types to assimilate the implications of the German data captured at the end of the war. These data had much to offer in the fields of high-speed aerodynamics and axial flow turbojet design, and allowed the Americans and Soviets to advance to a considerably more advanced second generation of jet-powered fighters far more quickly and successfully than would otherwise have been the case.

The first of these swept-wing fighters was the North American F-86 Sabre, a radical development of the FJ Fury naval fighter with flying surfaces swept at 35° in light of German research data. The Sabre fully met the US Air Force's short-term needs and, because of delays in the USSR's strategic heavy bomber programme, the service's medium-term requirements when it was evolved into a more capable type with air-search radar and a primary armament of devastating air-to-air unguided rockets fired as a salvo.

In the longer term the USAF needed a more advanced interceptor with a capable fire-control system used with air-to-air guided missiles. First off the mark in an effort to create such a fighter was Convair with its Model 7 submission, which was ordered in prototype form as the XF-92 to be powered by a 1600-lb (726-kg) thrust Westinghouse 130 turbojet supplemented for take-off and combat by a 6000-lb (2722-kg) thrust Reaction Motors LR-11 liquid-propellant rocket. This was a tailless delta type with mixed powerplant, and its wing was designed with the aid of Dr Alexander Lippisch, the aerodynamicist whose work had led to Germany's revolutionary Messerschmitt Me 163 rocket-powered fighter in World War II. Validation of the basic design's 60° swept delta wing (complete with full-span elevons for pitch and roll control) was entrusted to a smaller-scale aeroplane, the Model 7-002, which used components from five other aircraft and was powered by an Allison

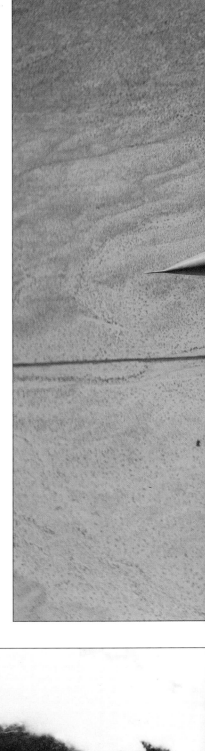

ABOVE. The Lockheed F-90 was a fairly radical development of the F-80 Shooting Star's concept, with finer fuselage lines and modestly swept flying surfaces, in an attempt to provide the US Air Force with a long-range penetration fighter.

LEFT. The North American F-86 Sabre was one of the great fighters of all time but, despite its swept flying surfaces, had been conceived originally with straight surfaces.

J33-A-23 turbojet. The Model 7-002 was generally successful, and development of the XF-92 proceeded without undue problems. In June 1949 the XF-92 was cancelled, but the Model 7-002 was kept in development as the XF-92A high-speed research aeroplane. Re-engined with the 6800-lb (3084-kg) J33-A-29 afterburning turbojet, the XF-92A eventually reached a speed of Mach 0.95 at 40,000-ft (12,190-m). Meanwhile the USAF was completing the competition for the fire-control system originally specified for the XF-92, and this was later installed in the modestly supersonic F-102 Delta Dagger interceptor that was evolved as the Convair Model 8 on the aerodynamic basis of the XF-92A with the advanced Pratt & Whitney J57 afterburning turbojet.

The design evolved to compete with the XF-92 was the Republic XF-91 'Thunderceptor'. This was a highly unusual type clearly based on the F-84F Thunderstreak swept-wing derivative of the F-84 Thunderjet straight-winged tactical fighter. In its fuselage and tail unit the XF-91 was conventional, but the wings and main landing gear units were decidedly unconventional in an effort to avoid the problems associated with low-speed wingtip stall in swept-wing aircraft. The wing was of the variable-incidence type to permit a higher angle of attack for take-off and landing, but this feature was combined with a planform of inverse taper and thickness, together with leading-edge slots. Thus the chord and thickness of the wing increased from root to tip, producing more lift at the tip than at the root, and this arrangement dictated that the main landing gear units be arranged to retract outward into the thick tips rather than inward into the thin roots.

Powered by a 5200-lb (2359-kg) thrust General Electric J47-GE-3 turbojet and a Reaction Motors LR-11 rocket, the latter using four nozzles arranged in pairs above and below the jetpipe, the XF-91 first flew in May 1949. In December 1952 the type exceeded Mach 1, and was later fitted with a butterfly tail whose V-angled surfaces replaced the conventional empennage. No production order was placed, but the XF-91 was extensively used for research before it was retired.

In 1946 the new Strategic Air Command of the USAAF placed an order for the development of a so-called penetration fighter, otherwise an escort fighter for the Convair B-36 strategic heavy bomber. The need for such an aeroplane had become clear over Germany in 1943 and 1944, when Boeing B-17 and Consolidated B-24 heavy bombers suffered considerably at the hands of German fighters until escort was provided by such classic types as the Republic P-47 and then by the great North American P-51. What was now demanded was a long-range fighter able to fly ahead of the bomber force and sweep aside all fighter opposition. Lockheed's Model 153 submission was thought to offer great potential and was ordered in the form of two XF-90 prototypes. The design had a number of similarities to the company's F-80 Shooting Star, but was of more advanced aerodynamic concept. It had a finely tapered forward fuselage, two laterally mounted 4200-lb (1905-kg) afterburning thrust Westinghouse J34-WE-11 turbojets, and flying surfaces swept at 35°. A radius of about 1100 miles (1770-km) was provided by considerable internal fuel supplemented by jettisonable wingtip tanks. This was calculated to provide an escort capability into the western USSR from bases in West Germany, and a potent offensive punch was provided by a combination of four 20-mm cannon with six 0.5-in (12.7-mm) machine-guns. The first aeroplane flew in June 1949, and was immediately revealed to be drastically underpowered. The USAF's requirement was changing at this time, so the project was cancelled.

Produced to meet the same basic need was another fighter with 35° swept flying surfaces, the McDonnell XF-88. This was powered by a pair of 3000-lb (1361-kg) thrust Westinghouse J34-WE-13 turbojets

Another penetration fighter prototype was the McDonnell XF-88, and though it failed to secure a production contract for not satisfying its hopelessly optimistic requirement, it ultimately paved the way for the great F-101 Voodoo fighter and reconnaissance aeroplane.

exhausting short and therefore efficient jetpipes under the trailing edge of the wings. This forced the designers to adopt an upswept arrangement for the tail unit, which became a feature of several later designs from the same stable. The first of two prototypes flew in October 1948, and the second was powered by J34-WE-22 engines with short afterburners to provide 3600-lb (1633-kg) thrust in combat. Speed was considered adequate, but as both range and ceiling were well below the required figures the programme was cancelled in 1950. The first prototype was then recast as the XF-88B testbed for the Allison XT38 turboprop, with which it undertook many flights from April 1953 with 27 different propellers featuring varying numbers of blades in diameters between 4 and 10 ft (1.2 and 3.05 m). The XF-88's airframe then went forward to provide the structural and aerodynamic basis for the great F-101 interceptor and reconnaissance fighter, a genuinely supersonic type.

As the XF-8 and XF-90 were about to be cancelled, the USA became caught up in the Korean War. Here the USAF's Boeing B-29 heavy bombers were used on interdiction missions designed to dry up the flow of supplies and reinforcements to the communist forces, but began to suffer at the hands of Mikoyan-Gurevich MiG-15 fighters in the process. This reopened the question of penetration fighters, spurring development of the F-101 with a lengthened fuselage to accommodate two Pratt & Whitney J57 afterburning turbojets.

The basically similar need in World War II had led McDonnell to propose an extraordinary escort fighter. This 'parasite' (perhaps more properly 'symbiote') fighter had to be small enough to be carried aloft by a bomber and released as required. As early as 1942 McDonnell had proposed such a type as the MX-472 for semi-external carriage by the B-29, and further evolution of the concept resulted during 1945 in four variants of the Model 27 design for internal carriage by the Northrop B-35 and Convair B-36 heavy bombers. Under the spur of early 'Cold War' events, McDonnell reworked the Model 27 into the XF-85 Goblin, which has been aptly described as 'a large egg fitted with flying surfaces'. The fuselage was virtually filled by the 3000-lb (1361-kg) thrust Westinghouse J34-WE-7 turbojet and its fuel plus the four 0.5-in (12.7-mm) machine-guns and their ammunition, the pilot being seated astride the engine under a bubble canopy immediately aft of the hook that permitted aerial release and recovery of this tiny fighter. The wings were swept at 37° and arranged to unfold from their vertically stowed position, which gave the fighter a 'hangared' width of only 5 ft 4.75 in (1.64 m), for an extended span of 21 ft 1.5 in (6.54 m)

ABOVE. The Lockheed F-94 Starfire was a straightforward evolution of the F-80 Shooting Star with an afterburning engine, radar (requiring a second member of the crew) and revised armament to produce an all-weather fighter.

after being lowered from the parent bomber but before release. The tail unit was quite remarkable, being an arrangement of six surfaces designed to provide adequate longitudinal and directional control yet fit inside the bomb bay of the B-36 without folding. The first of two XF-85s was flown in August 1948, but trials revealed that the type possessed dismal handling characteristics and poor performance, and that pilots found it extremely difficult to hook back onto the parent bomber. The order for 30 production aircraft had already been cancelled during 1947.

In the period after World War II the USAAF (and then the USAF) was similar to other major air arms in thinking of the day and night/adverse-weather fight roles as being the provinces of different types: the day fighter was conceived as a lighter single-seat type while the night/adverse-weather fighter was seen as a heavier two-seat type with radar. The USAAF's standard night fighter and intruder at the end of World War II was the mighty Northrop P-61 Black Widow, but consideration was soon given to a turbojet-powered successor. Some thought was given to a two-seat version of the Bell XP-83, a long-range single-seater that had flown in February 1945, but the performance of this early type was soon perceived to be inadequate.

Three companies produced prototypes meeting the USAF's requirement for a radar-carrying aeroplane with an armament of cannon or heavy machine-guns, a speed of at least 600 mph (966 km/h) and a ceiling of 40,000 ft (12,190 m). Two of these types led to production aircraft of increasingly potent capabilities (the Lockheed F-94 Starfire and the Northrop F-89 Scorpion), while the third was the Curtiss-Wright XP-87, which has the unfortunate distinction of being the last type produced by this illustrious company. The design featured four 20-mm cannon in a fixed nose installation and provision for four 0.5-in (12.7-mm) machine-guns in a dorsal barbette. Power was provided by four turbojets located in pairs on each wing: the first aeroplane was known as the Nighthawk and powered by four Westinghouse J34-WE-7 engines, while the second was known as the Blackhawk and powered by four General Electric J47-

GE-15 engines. the XP-87 Nighthawk first flew in March 1948, but the XP-87A Blackhawk never flew and an order for 88 J47-powered production aircraft was cancelled to provide additional funding for the F-89 and F-94 programmes.

The US Navy was also concerned to equip its carrierborne fighter squadrons with aircraft more effective than first-generation types such as the Ryan FR-1 Fireball with a composite powerplant (nose-mounted piston engine for long-range cruise and tail-mounted turbojet for high performance in combat) and the Mcdonnell FH-1 Phantom with two small Westinghouse J30 turbojets. Though both these types served with operational squadrons, the type that may be regarded as the navy's first genuinely operational turbojet-engined fighter was the North American FJ-1 Fury, the straight-winged aeroplane from which the swept-wing F-86 Sabre was then evolved for the air force. By a neat reversal of the process the navy then took a navalized version of the F-86E as the FH-2 Fury.

From the Fireball Ryan developed the XF2R with the General Electric XT31-GE=2 turboprop in place of the FR-1's composite powerplant. The XFR2R first flew in November 1946 but was not ordered into production. The navy had been sufficiently impressed with the concept of the composite powerplant, however, to order a further type in prototype form. Three Curtiss-Wright XF15C prototypes were ordered in April 1944. It had one Pratt & Whitney R-2800-34W piston engine in the nose delivering 2100 hp (1566 kW) to a four-bladed propeller and a de Havilland H.1B turbojet in the central fuselage delivering 2700-lb (1225-kg) thrust via a long jetpipe exhausting under the tail. The composite powerplant could provide the necessary performance, for the XF15C was capable of 469 mph (755 km/h) and 1385 miles (2229 km), but the type had unfortunate

BELOW. More akin to the original straight-wing concept for the Sabre land-based fighter was the North American FJ Fury naval fighter, seen here in the form of its XFJ-1 prototype. Design work was launched in 1944.

handling deficiencies and was not ordered into production.

Viewed with hindsight the concept of the composite powerplant appears a trifle bizarre, but a considerably greater oddity was another experimental naval fighter of the period. This was the Vought XF5U based on the V-173 aerodynamic research aeroplane. The XF5U had a basically circular wing with twin vertical surfaces at its rear 'corners' outside two stability flaps and inside two projecting 'ailavators' for pitch and roll control. The primary structure was of Metalite, a material of bonded aluminium and balsa that offered exceptional strength with great lightness. The powerplant comprised two 1600-hp (1193-kW) Pratt & Whitney R-2000-7 radials buried in the thick inner portions of the wings and driving, via gears, two large four-blade propellers located on the forward 'corners' of the wing. the V-173 confirmed that the type offered viceless handling characteristics as well as an exceptional speed range between 20 mph (32 km/h) and 460 mph (740 km/h). The prototype was completed in August 1945, but it was 1947 before the special propellers were ready. Flight trials were scheduled for 1948, but before these could be started the programme was cancelled and the sole

LEFT. One of the great disappointments in aviation history must be the decision to cancel the 'circular wing' Vought F5U naval fighter just before the XF5U-1 prototype flew.

LEFT. An early naval fighter that progressed only marginally past the prototype stage was the Vought F6U-1 Pirate (closest to the camera), which was soon overtaken by more advanced types.

BELOW. The Vought V-173 validated the basic aerodynamic concept for the F5U naval fighter, and proved to have excellent performance (within the limitations imposed by its low power) as well as truly viceless handling, including great agility.

XF5U destroyed — although only with great difficulty because of its very strong structure.

Development of more modern turbojet-powered fighters was proceeding during this period. Although the FJ-1 Fury became the navy's first fully operational jet fighter, it had in fact been beaten into the air by a competitor. The Vought XF6U Pirate first flew in October 1946 just seven weeks before the Fury prototype. The Pirate was conceptually a less advanced aeroplane, although it did provide its pilot with excellent fields of vision from a bubble canopy very close to the nose. This cockpit position was made possible by the use of wing-root inlets for the 3000-lb (1361-kg) thrust Westinghouse J34-WE-22 turbojet. Additional power was provided later by the 4200-lb (1905-kg) thrust J34-WE-30A engine, and this improved performance to the level at which 30 production aircraft were ordered. It took 18 months for these to reach the navy, and in this time the pace of development had been such that more modern types such as the Grumman F9F Panther and McDonnell F2H Banshee

were offering much improved performance.

The Panther and Banshee were the navy's fighter mainstays in the Korean War of the early 1950s, but by this time the implications of German high-speed flight research had been digested and incorporated into a new generation of fighter aircraft offering a transonic and finally a limited supersonic capability. These impressive fighters were the Vought F7U Cutlass swept flying wing design with a large central nacelle and two substantial vertical tail surfaces; the Douglas F3D Skynight straight-winged night interceptor that was also proposed in a swept-wing form as the F3D-3 for higher performance; the Douglas F4D Skyray tailless delta offering supersonic performance; the McDonnell F3H Demon swept-wing interceptor; the Grumman F9F Cougar swept-wing derivative of the straight-winged F9F Panther, and the North American FJ-2 Fury navalized version of the land-based Sabre.

These were single-seat aircraft optimized for the day role, but the navy was now concerned to provide its carriers with night and all-weather defensive capability using more advanced fighters carrying fire-control radar and air-to-air missiles. The first type to meet this need was the F5D Skylancer, which was basically an enlarged version of the Skyray with a fuselage of increased fineness ratio and a wing of reduced thickness/chord ratio for higher supersonic performance on the power provided by a Pratt & Whitney J57-P-8 afterburning turbojet. The first of four XF5D prototypes flew in April 1956 and went supersonic on its initial flight. The type proved capable of Mach 1.5, but the production version (with radar and four missiles in addition to an inbuilt armament of four 20-mm cannon) was cancelled in favour of more advanced fighters currently under development with full all-weather capability.

While these types were being evolved as actual or potential service aircraft, the navy was also pressing ahead with a pair of experimental fighters reflecting its concern about the take-off/landing requirements of such advanced jet fighters. These required the powerful boost of a steam catapult for take-off, and even with a system of arrester wires

RIGHT. It is remarkable in aviation how individual design concepts run through a string of designs from the same basic stable, and this McDonnell F3H Demon displays features, such as the location of the empennage on a boom projecting above the jet nozzle and the planform of the wing, still discernible in such later fighters as the F-4 Phantom II and even in the F-15 Eagle.

ABOVE. After the F6U Pirate, Vought's next naval fighter was the altogether more ambitious F7U Cutlass in which the short fuselage supported the powerplant and the wings. Running aft from each wing was a short boom ending in a vertical tail surface. There was no tailplane, the wings being fitted with elevon surfaces that operated differentially as ailerons and in concert as elevators.

RIGHT. The single most important concept with which designers had to come to grips in the late 1940s and early 1950s was the swept wing. Such a wing offered very considerable advantages in maximum speed and high-speed controlability, and it was often worth reworking a straight-winged design to incorporate a swept wing. Typical of this process was the US Navy's Grumman F9F Cougar, the simply evolved but considerably more effective swept-wing derivative of the F9F Panther.

LEFT. Another naval fighter that was recast with swept wings was the North American FJ Fury. This straight-winged design had provided a starting point for the F-86 Sabre swept-wing fighter for the US Air Force, and the wheel came full circle with the US Navy's adoption of a navalized F-86E as the FJ-2 Fury. This is a good example of the definitive FJ-4 version.

required a sizeable area for landing. The introduction of the angled flightdeck had eased matters by allowing the incorporation of separate take-off and landing areas, but even so the needs of still heavier aircraft were making larger and therefore more vulnerable aircraft-carriers increasingly the norm. Quite rightly, the navy appreciated that vertical take-off and landing (VTOL), especially for fighters, would reduce the scale of this problem, and from 1949 actively pursued a policy of VTOL research.

The fruits of this programme were two remarkable prototypes, the Lockheed XFV-1 and the Convair XFY-1 Pogo. Both aircraft were of the 'tail-sitter' concept and powered by the 5500-shp (4101-kW) Allison T40-A-6 turboprop driving large contra-rotating propeller units. These provided more thrust than the weight of the aircraft, making possible VTOL operation. The XVF-1 was the more conventionally configured of the two types, with a mid-set wing of low aspect ratio, but for VTOL capability had a cruciform arrangement of tail surfaces indexed at 45° to the wings and each fitted with a small castoring wheel on the outboard end of its trailing edge. For flight trials with an engine not cleared for VTOL operation the type was fitted with a lightweight but very stalky fixed landing gear arrangement to permit conventional rolling take-off and landing, and in this guise first flew in June 1954. The aeroplane flew 22 times, in the process recording 32 operations in the vertical mode, when variation of the engine power made possible descending,

BELOW. Apart from Convair, the other main exponent of the delta-winged planform for military aircraft in the USA was Douglas, which produced a number of notable types including the Skylancer interceptor, seen here in the form of XF5D-1 prototypes developed from the F4D Skyray.

ABOVE AND LEFT. One of the most remarkable fighters developed for the US Navy, and indeed for any air arm, was the Convair XFY-1 Pogo, a tail-sitter VTOL type that was ultimately used only for research work without any real thought to a production model.

hovering and ascending flight. No pure VTOL operations were undertaken with the only one of th two XFV-1s that flew.

The XFY-1 was a swept delta design with large dorsal and ventral fins, all four surfaces of this cruciform arrangement carrying castoring wheels to support the aeroplane vertically on the ground. This arrangement of flying surfaces made it impossible to fit the XFY-1 with the same type of conventional landing gear as the XFV-1, so Convair received the only example of the T40 turboprop cleared for VTOL operation. The Pogo made its first free VTOL flight in August 1954, and the flight test programme was very successful in overall terms, but like that of the XFV-1 confirmed that exceptional piloting was necessary for such a tail-sitter type.

The US Navy also investigated two other means to reduce the flightdeck requirements imposed on its new aircraft-carriers, namely the waterplane fighter and the variable-geometry fighter. The waterplane fighter required no flight deck at all, and was therefore attractive for a number of tactical reasons. The navy stressed the importance of performance at least comparable with that of contemporary landplane fighters, and this immediately dictated an airframe radically different from all previous waterplane fighters, which had been either structurally complex and high-drag floatplanes or flying boats with a large hull to provide the volume required for waterborne buoyancy.

The company selected to produce a waterplane interceptor was Convair, and the resulting XF2Y Sea Dart was a fascinating type. The company was instructed to investigate the 'blended hull' concept, in which the hull rode so low in the water that the wings provided part of the required buoyancy until the aeroplane had accelerated sufficiently to rise onto a planing bottom, while the National Advisory Committee for Aeronautics (from 1958 the National Aeronautics and Space Administration) investigated the hydro-ski concept. This latter was developed as an alternative to floats: when extended, the hydro-skis lifted the accelerating aeroplane onto the surface of the water, where the skis planed and so allowed the aeroplane to reach take-off speed; and when retracted they formed the lower surface of the fuselage without creating any drag.

For the Sea Dart Convair finally opted to combine the two concepts, and the aeroplane that emerged from the design phase was a small but elegant type reminiscent of the company's other delta-winged aircraft, except that the inlets for its two 3400-lb (1542-kg) thrust Westinghouse J34-WE-32 turbojets located in the dorsal position where they would be shielded from spray. During test runs there was considerable vibration and pounding from the two hydro-skis, and these were replaced by a single V-shaped ski. The Sea Dart first flew in April 1953, but continued problems with the hydro-ski, lack of power and a number of aerodynamic problems finally led to the cancellation of the whole programme in 1955.

Greater success attended the development of variable-geometry wings, which allow the use of a minimum-sweep position for take-off and landing, an intermediate-sweep position for fuel-economical cruise, and a maximum-sweep position for high dash performance. The navy's first essay in this field ws the Grumman XF10F Jaguar, which was conceived in 1948 as a possible successor to the F9F Panther. The aerodynamic features of a variable-geometry type had already been explored in a number of aircraft, most notably the Westland Pterodactyl IV developed in the UK during the early 1930s, the Messerschmitt P.1011 designed in Germany during World War II, and the experimental Bell X-5 sponsored by NACA and the USAF with the ultimate purpose of validating a fighter-type wing with sweep variable between 20° and 60°. The X-5 had not flown when Grumman set to work on the Jaguar, but a considerable quantity of engineering data was available and this proved valuable to the Grumman engineers, who were faced with a host of problems.

Originally the XF10F had been planned as a development of the Panther with clipped delta flying surfaces. The concept was then refined to the point at which a tilting variable-incidence wing was adopted for reduction of the take-off and landing speeds. The navy then added additional responsibilities to the basic fighter role, and these so increased the structure weight that the company came up with the notion of providing a variable-geometry wing. The first variable-geometry scheme envisaged a

reduced landing speed from 132·5 mph (213 km/h) to 109 mph (175 km/h). The Jaguar was also provided with an advanced control system that included a delta surface forward of the fin to act as a servo for the all-moving tailplane and so improve control response at transonic speed.

The engineering of so complex an aeroplane took considerably longer than expected, and it was May 1953 before the XF10F prototype flew, three years behind schedule. The novel servo-control system for the tailplane proved far too slow in operation, and was replaced by a

wing with only two positions (a sweep of 13° 12′ for take-off and landing, and a sweep of 42° 30′ for all other portions of the flight envelope) plus a mechanism to move the wing roots forward and so maintain the correct relationship between the centres of gravity and lift as the wings were swept aft. Further refinement was added after this when it was decided to allow the wing to be swept at any angle between the minimum- and maximum-sweep angles. The provision of such wings increased the type's weight by 2200 lb (998 kg) but

conventional powered tailplane, but it soon became clear that considerable revision of the basic design would have to be undertaken before production aircraft could be considered. Grumman therefore ended the programme. It is worth noting that the type's most revolutionary feature, the variable-geometry wing planform, never gave any trouble and was perhaps the most successful single component of the whole design, fulfilling all the hopes entertained for the concept.

By the time of the Jaguar's demise the navy was firmly committed to the

From its successful F-8 Crusader fighter, the Vought company developed the F8U-3 Crusader III retaining the original model's variable-incidence wing but using a more powerful engine aspirated via a forward-raked chin inlet. In preference to this type, however, the US Navy selected the McDonnell F4H that eventually became the F-4 Phantom II.

supersonic fighter, and types that entered service were the Grumman F11F Tiger (one of the world's first warplanes with an area-ruled fuselage) and Vought F8U Crusader. The latter may be regarded as the navy's counterpart to the USAF's North American F-100 Super Sabre, but was in overall terms a more capable fighter with the distinctive feature of a variable-incidence wing to reduce take-off and landing speeds while keeping the fuselage comparatively level so that the pilot's view of the flightdeck was not impaired.

So successful was the Crusader, moreover, that a serious effort was made to create a Mach 2 development as the F8U-3 Crusader III. This bore a strong external resemblance to the baseline Crusader, but was virtually a new aeroplane characterized by the revised forward fuselage (with a pointed nosecone and forward-raked 'sugar scoop' inlet) and higher aspect ratio ventral fins that were angled down from the horizontal position for additional stability in supersonic flight. The type first flew in June 1958 and proved to have excellent performance, but the competing Mcdonnell F4H Phantom II was preferred for production.

The only other naval fighter to have reached the hardware stage and then failed to secure production orders was the General Dynamics/Grumman F-111B, the naval version of the F-111 tactical fighter. This whole project was schemed around the political and financial desirability of a single airframe/powerplant package to satisfy significantly different air force and navy requirements. The basic variable-geometry aeroplane was designed by General Dynamics, with Grumman

ABOVE. The F-86 Sabre paved the way for the Western world's first genuinely supersonic fighter, the North American F-100 Super Sabre with more highly swept flying surfaces, considerably more power from an afterburning turbojet, and a number of aerodynamic and system refinements reflecting the US aero industry's great experience of transonic and supersonic flight with a number of service and experimental aircraft.

RIGHT. Another fighter that profited from experience with transonic service and supersonic experimental aircraft was the McDonnell F-101 Voodoo fighter, which was evolved somewhat radically from the XF-88 penetration fighter prototype.

BELOW. The Republic F-105 Thunderchief was another 'Century series' tactical warplane that easily secured supersonic performance through the use of great power combined with an aerodynamically elegant airframe, in this instance with a waisted or 'Coke-bottle' fuselage to avoid the drag penalties of great variations in cross section.

primarily responsible for the naval version. This first flew in May 1965, and was plagued by a number of weight and system problems. the F-111B was cancelled in 1968, although Grumman used its experience with the type's variable-geometry wings (as well as its radar fire-control system and mighty Phoenix air-to-air missiles) in its design for the F-111B's successor, the F-14 Tomcat fleet-defence fighter.

Two other navy projects worthy of note, even though they did not reach the hardware stage, are the Douglas F6D Missileer and the Grumman XF12F. The Missileer appeared to be an anachronism, for in the era of Mach 2 fighters with swept flying surfaces it was designed as a firmly subsonic type with straight wings and a powerplant of two Pratt & Whitney TF30-P-2 turbofans for long endurance. As its name suggests, the Missileer was

RIGHT AND FAR RIGHT. The Lockheed SR-71 'Blackbird' was phased out of first-line service in 1990, but remains one of the greatest technical achievements in the history of powered flight. It is unlikely that the SR-71's absolute records for speed and sustained ceiling will be beaten for many years.

not in itself a fighter but rather a radar-equipped launch centre for six AAM-N-10 Eagle long-range air-to-air missiles, which were regarded as the interceptors. Production of 120 aircraft was ordered, but the complete project was cancelled in 1961 before work on the first prototype had been started.

The XF12F was proposed for the requirement won by the Phantom II. The design had clear affinities to that of the F11F, but was for a larger two-seat aeroplane powered, in prototype form, by two 15,600-lb (7076-kg) afterburning thrust General Electric J79-GE-3 turbojets supplemented by a throttlable

whose first genuinely supersonic fighter was the F-100 Super Sabre, which began life as a reworking of the Sabre with its wings swept at 45° but then became a radically different aeroplane. The F-100 was the first of the USAF's 'century series' of supersonic fighters, and was followed by production types such as the McDonnell F-101 Voodoo interceptor and reconnaissance aeroplane, the Convair F-102 Delta Dagger radar-carrying interceptor, the Lockheed F-104 Starfighter clear-weather interceptor, the Republic F-105 Thunderchief strike fighter and the Convair F-106 Delta Dart upgraded development of the F-102.

The gaps in this sequence mark a series of remarkable aircraft. The Republic XF-103 was a projected Mach 3 interceptor powered by a composite powerplant: one 22,100-lb (10,025-kg) thrust Wright YJ67-W-3 turbojet and one 37,400-lb (16,965-kg) thrust Wright XRJ55-W-1 rocket engine. The original design called for a flush cockpit with a periscope to give the pilot a degree of vision, and the armament envisaged was six GAR-1 Falcon missiles on retractable launchers in fuselage side bays and 36 2.75-in (70-mm) unguided rockets, to be replaced in later aicraft by two GAR-1 and two nuclear-tipped GAR-3 Falcon missiles.

5000-lb (2268-kg) thrust rocket engine. The crew was to be located in a jettisonable escape capsule, and Mach 2+ performance was envisaged. Two prototypes were ordered in 1956, but these were later cancelled.

The navy's push into the supersonic age was matched by that of the air force,

The North American YF-107 was originally designated the YF-100B, and as its original designation suggests it was a derivative of the Super Sabre. The type was intended as an interceptor and fighter-bomber, and three aircraft were completed. The first of these flew in September 1956, revealing a radical revision of the forward fuselage with a pointed nosecone, requiring aspiration of the 24,500-lb (11,115-kg) afterburning thrust Pratt & Whitney J75-P-9 via a large dorsal arrangement with bifurcated inlets just behind the cockpit. Speeds of mach 2.2+ were attained in the flight test programme, but development was ended in 1957 for financial reasons.

BELOW. In the General Dynamics YF-16 the US Air Force moved away from the massive Mach 2 fighters that had proved somewhat limited in the Vietnam War, and instead shifted to a smaller but aerodynamically more sophisticated type with slightly lower performance but considerably greater versatility, agility and overall maintainability.

The North American YF-108 Rapier was to have been a Mach 3 interceptor with missile armament, tandem seats for the two crew, a canard layout and a powerplant comprising two General Electric J93 turbojets. The project was cancelled in 1959.

The designation F-109 was allocated to a planned interceptor development of the Ryan X-13 Vertijet VTOL research aeroplane, but the two prototypes ordered in Fiscal Years 1959 and 1960 were cancelled. F110 was the original USAF designation for a land-based derivative of the Phanton II naval fighter, but this finally entered service as the F-4 after the rationalization of the US

RIGHT. The Northrop YF-17 was the losing competitor to the General Dynamics YF-16 in the US Air Force's Light-Weight Fighter competition, but was then extensively developed, enlarged and improved by McDonnell Douglas and Northrop to become the US Navy's F/A-18 Hornet dual-role fighter and attack aeroplane.

ROCKWELL/MBB X-31A

The X-31A was designed as the manned aeroplane component of the Enhanced Fighter Maneuverability programme, and as a result of political pressure to increase co-operation between the countries of the NATO alliance in basic research, became the first aeroplane in the X-series to be designed with the aid of a foreign aerospace company, in this case Messerschmitt-Bolkow-Blohm of West Germany.

The programme's Phase I feasibility study came to the conclusion that close-in air combat would in all probability remain a feature of future fighter operations, and that enhanced agility would therefore be useful for any future fighters. The X-31A was therefore designed to break the 'stall barrier', and thus provide close-in combat agility above the normal stall angles of attack. Phase II of the programme, which is managed by the US Defense Advanced Research Projects Agency together with the West German ministry of defence, from September 1986 evolved the design for a fighter-configured aeroplane with Rockwell responsible for the overall structural and aerodynamic design and MBB for the control systems and thrust-vectoring engine nozzle. The X-31A is intended to explore the integration of several technologies applicable to an expanded manoeuvring envelope using vectored thrust and integrated control systems to provide rapid target acquisition, sustained manoeuvre capability at speeds well below the conventional stalling speed, and accurate fuselage pointing for the engagement of future targets at all speeds between low and supersonic.

Considerable input for the design was provided by the two companies' earlier efforts in this field, notably the HiMAT remotely piloted vehicle and the TKF-90 fighter design. The programme includes two flight-test aircraft, and these feature a long fuselage complete with high-visibility cockpit enclosure, ventral inlet and aft-mounted engine with paddles in the exhaust plume for vectoring of the thrust. The wing is located about midway along the fuselage, and is a composite structure of the compound delta planform, while the surfaces toward the ends of the fuselage are the forward-mounted canards and the aft-mounted

vertical surface. A digital fly-by-wire system is used for control of the control surfaces and the vectoring system. The first of the aircraft should fly in 1990, and an ambitious flight test programme will probably yield very important results in the creation of the next generation of warplanes.

> ### SPECIFICATION
> ### Rockwell/MBB X-31A
>
> **Type:** enhanced fighter manoeuvrability demonstrator aeroplane
> **Accommodation:** one
> **Electronics and operational equipment:** communication and navigation equipment
> **Powerplant:** one 10,600-lb (4808-kg) thrust General Electric F404-GE-400 non-afterburning turbofan
> **Performance:** maximum speed 597 mph (961 km/h) or Mach 0.9 at 35,000 ft (10,670 m)
> **Weights:** empty 10,212 lb (4632 kg); maximum take-off 13,968 lb (6335 kg)
> **Dimensions:** span 23 ft 10 in (7.26 m); length 43 ft 4 in (13.21 m) excluding probe; height 14 ft 7 in (4.44 m); wing area 226.3 sq ft (21.02 m^2) and canard foreplane area 23.6 sq ft (2.19 m^2)

LEFT. A Grumman proposal of the late 1970s reveals how the designers of the period were considering low-cost aircraft with blended aerodynamics and forward canards rather than a tailplane, as well as thrust-vectoring engine nozzles for improved field performance and flight agility.

services' different designation systems into a single system in October 1962. The final aeroplane in the original F-series was the F-111 interdictor from General Dynamics.

Most of the early designations in the new F-series were reserved for existing aircraft, and the first new aeroplane in the sequence was the Northrop F-5 Freedom Fighter, a lightweight type intended mainly for export under American military aid programmes to friendly nations. The first wholly new type for US military service was the Lockheed YF-12A. The YF-12A paved the way for the almost legendary SR-71A 'Blackbird' strategic reconnaissance platform that was retired from first-line USAF service in 1989, but even so the whole programme is still shrouded in secrecy and uncertainties. The origins of the programme lay with the A-11, which was built at Lockheed's 'Skunk Works' and first flown in April 1962 as the precursor of an interceptor and reconnaissance family capable of Mach 3+ performance at high altitude. The type is based on advanced aerodynamics using a blended fuselage/wing design built largely of titanium alloys and covered in a special heat-radiating paint that led to the type's nickname. The powerplant comprised a pair of 32,500-lb (14,740-kg) afterburning thrust Pratt &Whitney J58 (JT11D-20B) bypass turbojets (or turbo-ramjets) which at high speeds produced their power not only as direct thrust from the exhaust nozzle but also as suction at the inlet.

The fighter derivative of the basic model was the experimental YF-12A, of

F-16XL

General Dynamics F-16XL
The cranked-arrow wing planform was considered for the original F-16 fighter, but was not adopted for a number of reasons. So promising was this wing layout, however, that General Dynamics later returned to it for the F-16XL technology demonstrator. The fuselage was stretched to accommodate the broader-chord wing, and fuel capacity was increased by 82 per cent, allowing the aeroplane to carry 'double the weapon load 45 per cent further', using no fewer than 17 hardpoints. In the air the F-16XL proved to possess phenomenal agility and higher performance than the basic F-16, and was also extremely reliable.

which at least four were produced with the A-11's original short fuselage, a Hughes pulse-Doppler fire-control system and, in the fuselage chine bays originally used for the carriage of reconnaissance equipment, four AIM-47A air-to-air missiles. The capabilities of the YF-12A were indicated by world straight-line and closed-circuit speed records, and a world record for sustained height. The YF-12A never served operationally, but was important in several evaluation programmes.

The only other publicly revealed designation in the new F-series of designations is YF-17, which was used for the Northrop Model P600 lightweight fighter than was evaluated against the YF-16 (General Dynamics Model 401) in the USAF's Light-Weight Fighter (LWF) competition of 1974. The YF-17 lost the contest, which resulted in contracts for the F-16 Fighting Falcon. The LWF competition resulted from the USAF's appreciation that in the Vietnam War its Mach 2 fighters had been reduced in overall capability by operating against lighter and more manoeuvrable fighters in an environment for which they had not been designed. The new fighters placed emphasis on greater reliability (and thus reduced maintenance) together with superior tactical flexibility and reduced outright performance in a smaller but more agile airframe using advanced aerodynamics, structures, electronics and a fly-by-wire control system.

Even though the YF-17 lost the LWF competition, it was taken up as the basis of the McDonnell Douglas/Northrop F/A-18A Hornet, currently the most important dual-role fighter and attack aeroplane in the inventory of the US Navy and US Marine Corps.

LEFT. Another Grumman concept of the late 1970s was a fighter optimized for fuel-economical cruise at supersonic speed by the use of blended aerodynamics, canards and thrust-vectoring engines located in the wing roots.

ABOVE. An artist's impression of the proposed General Dynamics F-16XL from the early 1980s.

SOVIET FIGHTERS

Yet again, the industry of US aeroplane manufacturers was matched and indeed excelled in fighter design on the other side of the 'Iron Curtain'. Here the design organizations that dominated the field were the Lavochkin, MiG (Mikoyan-Gurevich), Sukhoi and Yakovlev bureaux. Like the Americans, the Soviets had started development of turbojet-powered fighters before the end of World War II, but were then content to operate a number of interim types as they digested the research data they captured from the Germans. The Soviets also seized many aerodynamicists, engineers and engine designers as well as numbers of specialists in allied fields. This German influence soon became apparent in a number of Soviet aircraft and turbojet engines.

Lavochkin had been one of the three most important design bureaux for piston-engined fighters in World War II, but seemed never to come fully to grips with the different requirements of turbojet-engined fighters. The starting point for the Lavochkin effort, as for those of the MiG and the Sukhoi bureaux, was the Junkers Jumo 004B axial-flow turbojet, a German type taken over by the Soviets with the designation RD-10 for continued development in Kazan. The February 1945 Soviet requirement for a jet-powered fighter specified a single 1984-lb (900-kg) thrust RD-10, and the limited power of this engine dictated that the designers had to exercise considerable ingenuity in keeping down weight and optimizing the engine installation. The Lavochkin team's response was a small machine and a pod-and-boom fuselage allowing the use of an efficient straight-through design from the circular nose inlet to the nozzle under the boom. Five prototypes were built for trials from late 1946, but the type was so beset by problems including a high structure weight and sluggish performance that the programme was cancelled in 1947.

The same basic structural and aerodynamic features were improved for

BELOW. In the early days of turbojet power, designers were at great pains to minimize engine thrust loss as a result of a single engine exhausting at the tail via a long jet pipe. The solution adopted in many Soviet fighter prototypes, such as the straight-winged Lavochkin La-156, was a pod-and-boom fuselage design with the engine located in the deep forward section of the fuselage to exhaust under the cockpit.

LEFT. As designers began to appreciate and apply the lessons of swept-wing aerodynamics, engine designers were producing turbojets of greater sustained power, and this allowed the development of single-engined types such as this Lavochkin La-168, which was in effect the prototype for the La-15, an excellent fighter that failed to enter production as large-scale production of the Mikoyan-Gurevich MiG-15 had already been authorized.

the following La-152 and for the La-154 version of the La-152 with a revised wing. These flew in October 1946 and later in the same year, paving the way for the La-156 that was in essence the La-154 with an RD-10F afterburning engine delivering a maximum thrust of 2425 lb (1100 kg). The type first flew in September 1947, and although considerably better than preceeding Lavochkin jet fighter prototypes, was clearly not comparable with the swept-wing fighters that were now beginning to emerge. The final development of this straight-winged series was the La-174TK, which first flew in January 1948 as a research type for the investigation of very thin, straight wings as an alternative to swept wings in overcoming compressibility problems.

The Soviets' first swept-wing aeroplane was the La-160, which was basically the La-154/156 with a lengthened fuselage and 35° swept flying surfaces. The first of several prototypes flew in July 1947 and revealed much superior performance to the straight-winged Lavochkin types, but no production was contemplated as it was now recognized that the RD-10 engine and the pod-and-boom fuselage arrangement were obsolete. Thus the bureau produced the La-168 with the 5000-lb (2268-kg) thrust Rolls-Royce Nene I centrifugal-flow turbojet in a conventional fuselage. The engine was located toward the tail, where it exhausted through a short jetpipe, and was aspirated from a circular nose inlet via bifurcated ducting round the pressurized cockpit. The wing was swept at 37° 20′, and a T-tail was used to allow the location of the tailplane farther to the rear than would otherwise have been possible, improving longitudinal control. Flight trials were started in April 1948 and immediately confirmed that the La-168 was a fighter prototype of considerable peformance and potential. The MiG-15 had already been ordered into production, however, so no production of the La-168 was ordered.

Lavochkin then scaled down the basic design to produce the La-174D tailored round the smaller Rolls-Royce Derwent turbojet. This prototype first flew in August 1948, and performance was so

BELOW. The Lavochkin La-174TK was essentially the La-156 with a thinner wing. In fact, the 6 per cent thickness/chord ratio was the smallest in the world at that time, and yielded handsome performance benefits over the La-156. The La-174TK was intended only for research, however, and no production was ever planned.

good that an order was placed for 500 La-15 fighters with the Soviet version of the Derwent, the 3527-lb (1600-kg) thrust RD-500. The type proved popular and successful in service, and it is interesting to note that the La-15 had much the same performance as two British jet fighters, the Gloster Meteor with 100 per cent more power through the use of two Derwents, and the Hawker Hunter again with 100 per cent more power through the use of a single more modern turbojet.

The pace of progress in this period of intense but cold hostility between the USSR and USA was furious, and this is reflected in the appearance of the La-176 for a first flight in September 1948. This was a derivative of the La-168 with a wing swept at 45° (the first in the world) and several other aerodynamic refinements, and with a 5004-lb (2270-kg) DR-45 turbojet as powerplant. The La-176 was transonic in level flight and capable of Mach 1 in a shallow dive, but was again pipped to a production order by a rival MiG design, in this instance the MiG-17.

In October 1948 the Soviets issued a requirement for a genuinely transonic fighter, and the Lavochkin bureau responded with the La-190 prototype. This was an all-new design of considerably more advanced concept than previous Lavochkin fighters, and featured wings swept at 55° and a delta tailplane located about three-fifths of the way up the broad-chord vertical tail. Power was provided by a potentially important new Soviet axial-flow turbojet,

ABOVE. Another method of keeping the jetpipe as short as possible was to use a large and well-swept vertical tail with the tailplane located high on it and therefore well aft where it had an adequate leverage. The unusual nose of this Lavochkin La-200B results from the use of one chin and two cheeks inlets to leave the nose free for the radome over the search radar's antenna.

LEFT. The Mikoyan-Gurevich I-320 was the prototype for a proposed all-weather fighter, and the need to carry a weighty search radar led directly to this prototype's unusual lower-fuselage contours. Two engines were essential for adequate performance despite the weight and drag of the radar installation, and in this instance the engines were located as one under the fuselage exhausting in line with the wing roots, and one in the tail exhausting in the conventional manner. The I-320 remained a prototype.

ABOVE AND FOLLOWING PAGE. The classic Mikoyan-Gurevich MiG-21 was evolved to production standard only after an extensive prototype programme that saw flight trials with swept-wing and tailed delta design, the latter being adopted.

the Lyul'ka AL-5 delivering 11,023-lb (5000-kg) thrust, providing true transonic performance. Although planned originally as a day fighter, the prototype was completed in limited all-weather form with a radar radome in the upper portion of the nose inlet. Flight trials were undertaken from early 1951, and though the type was deemed generally satisfactory, some problems and the unreliability of the AL-5 led to the termination of the whole programme in favour of the MiG-19.

In January 1948 the bureau received details of a requirement for an all-weather interceptor, and again responded with a 'clean paper' design designated La-200 and powered by two examples of an engine derived from the RD-45, the 5004-lb (2270-kg) thrust Klimov VK-1. These were located in a tandem arrangement, with the forward engine exhausting from the bottom of the central fuselage via an S-shaped jetpipe and the aft engine through a nozzle at the tail, and were aspirated via an annular nose inlet round the centrebody that accommodated the antenna of the search radar. The flying surfaces were swept at 40°, and the crew of two was accommodated side-by-side under a large canopy. The first prototype flew in February 1950, and was in all performance respects far ahead of Western contemporaries. The range was (1243 miles (2000 km), but in November 1950 a range of 2175 miles (3500 km) was demanded, together with longer-range radar. The design was recast as the La-200B with greater fuel capacity and a larger antenna for the more powerful radar. This led to a redesign of the nose to accommodate the antenna; and the front engine was now aspirated via a chin inlet, while the rear engine was fed with air by two 'elephant ear' inlets projecting from the sides of the nose behind the radome. The La-200B first flew in July 1952, and while range was improved considerably, overall performance was reduced. The La-200B was beaten to a production order by the Yak-25.

The bureau's last design was another far-sighted fighter, planned in response to a January 1954 requirement for a super-interceptor able to cruise long distances at high altitudes with missiles able to provide a 'snap-down' attack capability against targets at lower altitudes. The La-250 was a massive type with substantial fuel capacity, two 14,330-lb (6500-kg) thrust Lyul'ka AL-7 turbojets in long lateral installations, and 57° swept flying surfaces that included a delta wing and slab delta tailplane. The first prototype was lost on its first flight in July 1956, and investigation revealed a severe roll-coupling problem derived from the combination of a long heavy fuselage and small wings. Enormous effort went into the creation of a new electronic flight-control system, and a successful first flight was recorded in spring 1957. The flight test programme was bedevilled by a number of accidents and engine problems, and the complete project was cancelled just before Semyon Lavochkin's death in 1960.

The MiG bureau had not enjoyed notable success with its piston-engined fighters during World War II, but was destined to come into its own with the advent of turbojet power. The bureau's first turbojet-powered machine was the I-300 prototype, a virtual pod-and-boom design that made the first flight by a Soviet jet aeroplane in April 1946. The two captured 1764-lb (800-kg) thrust BMW 003A engines (later built in the USSR with the designation RD-20) were located in the pod portion of the fuselage, and were aspirated via a bifurcated circular nose inlet to exhaust under the central fuselage under the wing trailing edge. In May 1947 the type was accepted for service as the MiG-9, which with the Yak-15 became the USSR's first operational jet fighter. A number of development models based on the MiG-9 were produced, but the bureau then went forward to the I-310 prototype for the MiG-15 series, the I-330 prototype for the MiG-17 series, and then the I-350 single-engined and I-360 twin-engined prototypes. The latter of these two resulted in the MiG-19 that was the USSR's first supersonic fighter. The MiG-

19 was built in several models, and was also used for a number of experimental purposes.

In the middle of this effort devoted to single-seat fighters intended for the clear-weather role, the bureau also found the design capacity for an all-weather fighter to meet an exacting air force requirement of January 1948 for an all-weather fighter with radar. The resulting I-320 prototype

flew in 1950, and was based aerodynamically on the MiG-15 although the fuselage was wide enough for side-by-side ejector seats, and the powerplant was a tandem arrangement of 5004-lb (2270-kg) thrust RD-45F turbojets. These were fed from a bifurcated nose inlet under the small radome, the forward engine exhausting under the wing trailing edge and the rear engine at the tail, which was clearly derived from that of the MiG-15. The I-320 was later re-engined with 5952-lb (2700-kg) thrust Klimov VK-1 turbojets, and though clearly superior to the La-200 was not seriously considered for production.

In 1953 the Soviet authorities issued a requirement for a Mach 2 clear-weather interceptor with limited ground-attack capability. At this time the USSR's

Central Aerodynamics and Hydrodynamics Institute had arrived at two basic configurations for aircraft of the required performance level. Both were based on a cylindrical fuselage with a swept all-moving tailplane and a wing in the low mid-set position, but the difference came in the wing itself. One was a conventional type with a leading-edge sweep of between 58° and 62°, and other a delta with 57° or 58° leading-edge sweep. The MiG bureau produced prototypes in both configurations. The original Ye-50 that flew in mid-1955 can be regarded as a pre-prototype, for in the absence of the planned Tumanskii R-11

This illustration of Mikoyan-Gurevich fighters in a Soviet air museum highlights the evolutionary nature of modern fighter development: from front to back the fighters on display are the MiG-17, MiG-19, MiG-21 and MiG-23.

turbojet it was fitted with an interim composite powerplant comprising an RD-9Ye afterburning turbojet and an S-155 rocket engine; the aeroplane had swept conventional wings based on those of the MiG-19. An RD-9Ye turbojet was also used in the Ye-4 with a delta wing, which first flew in December 1955.

The definitive prototypes with R-11 engines were the Ye-2A with swept wings and the Ye-5 with delta wings. these flew in May and June 1956 respectively, and were soon involved in comparative trials that showed the tailed delta configuration to have slight performance and operational advantages. The Ye-5 therefore became the basis for the Ye-6 prototype that was used to eradicate the various propulsion and flight-control problems besetting the programme, and in 1958 production was authorized of the MiG-21 fighter. The MiG-21 programme spawned its own series of developments through various prototypes, and there were also a number of experimental and record-breaking prototypes such as the Ye-33 version of the MiG-21U operational conversion trainer used for climb and altitude records for women pilots; the Ye-66 version of the MiG-21F used for a speed record; the Ye-66A with a U-2 rocket in a belly pack for an altitude record, the Ye-66B with twin rockets; the Ye-76 version of the MiG-21PF for a number of women's records; the Ye-8 with a powered canard foreplane to validate such a feature for the proposed MiG-21Sht attack fighter; the MiG-21DPD with two direct-lift jets in an extra fuselage bay on the centre of gravity, and the A-144 with a scaled-down version of the wing proposed for the Tupolev Tu-144 supersonic airliner.

The same basic aerodynamics were used in the Ye-150 family of high-speed research aircraft, which introduced features such as stainless steel and titanium into the airframe to withstand the heat at speeds of more than 1491 mph (2400 km/h). The ultimate expression of this philosophy was the Ye-166 used for exploration into the structural and aerodynamic requirements of flight at speeds of more than 1864 mph (3000 km/h).

Throughout the period in which it was developing the MiG-21, the MiG bureau was also involved in the design of large air-launched missiles yet it still had design capacity for another series of

aircraft. This was the I-3 series of fighter-bomber prototypes powered by the 18,518-lb (8400-kg) thrust Klimov VK-3 turbojet and having about double the empty weight of the initial MiG-21 variants. The first prototype of the series was the I-1 (otherwise known as the I-370) with a wing swept at 60°, and this first flew in November 1956. This led to several further developments epitomized by the I-3U (I-380) fighter-bomber beaten by the Su-22, and the I-3P (with an unknown alternative designation in the I-380 series) radar-equipped interceptor beaten by the Su-9. Further development of the basic concept, in this instance with the 20,500-lb (9300-kg) thrust Lyul'ka AL-7F, resulted in the I-7K that first flew in January 1959 and proved capable of flight at Mach 2.35. Evolution via I-7D, I-7P and I-7U prototypes led to the I-75F all-weather interceptor, but this too was not ordered into production.

Although extremely high flight performance was still required of Soviet warplanes operating in a number of specialized roles, there was increasing concern in the Soviet tactical air arms about the increased field requirements of high-speed aircraft with conventional

BELOW AND RIGHT. The Mikoyan-Gurevich MiG-23 fighter was developed as successor to the MiG-21 with greater range and, perhaps more importantly, the ability to operate from shorter runways. The definitive MiG-23, with its swept flying surfaces, resulted from the Ye-231 variable-geometry prototype, but serious consideration was also given to a production derivative of the Ye-230 tailed-delta prototype with two vertically mounted lift jets positioned on the centre of gravity.

configurations, and about the poor payload/range performance of these machines. The MiG bureau was planning a MiG-23 successor to the MiG-21, and the factor that caused the bureau the greater worry was the length of the runway that the new fighter would require. The two remedial approaches were direct lift and variable-geometry wings.

Trials in the direct-lift approach started with the MiG-21DPD to validate the basic concept, but were then taken a stage further with the Ye-230 prototype. This was built in parallel with the Ye-231 variable-geometry prototype to ensure maximum commonality for any production type resulting from the twin programmes. The Ye-230 was of tailed delta configuration, was powered by a Lyul'ka AL-7F-1 afterburning turbojet, and had the same type of lift engine arrangement as the MiG-21DPD, namely two turbojets (probably Koliesov units) located vertically on the centre of gravity with air drawn from above past a rear-hinged louvred dorsal door and exhausted downward through a grid of ventral transverse louvres which could be angled by the pilot to provide a forward thrust component during transition to forward flight.

The Ye-231 variable-geometry prototype was almost identical to the Ye-230 apart from its lack of lift jets and the use of variable-geometry swept wings similar to those of the General Dynamics F-111, the first operational variable-geometry warplane. Comparative trials revealed the superiority of the variable-geometry arrangement, and the Ye-231 thus became the precursor of the MiG-23 fighter, later adapted with a modified nose and simpler engine arrangements as the MiG-27 attack aeroplane.

The threat apparently posed by the North American XB-70 Valkyrie Mach 3 bomber was taken so seriously by the Soviets that in 1958 the MiG bureau was ordered to produce an interceptor able to tackle the American bomber. The bureau was instructed to ignore virtually every aspect of flight performance but outright speed, rate of climb and service ceiling in an airframe that was to be developed quickly by the use of existing technologies. This removed the possibility of delays resulting from slippage in the development of new technologies, and helped to ensure that the interceptor would be available at the time of the B-70's proposed service debut in 1964. The bureau chose a nickel-steel alloy as the primary airframe material, with titanium alloy leading edges. The Ye-266 prototype first flew in about 1964, and its airframe design was conceptually akin to that of the North American A-5 Vigilante naval attack aeroplane. It featured a large fuselage (comprising mainly the powerplant arrangement of two Tumanskii R-31 afterburning turbojets plus their variable-geometry inlets and fully variable nozzles), high-set wings of broad chord and a modest sweep of 40° declining to 38° outboard of the outer pylon, slab tailplane halves and outward-canted vertical tail surfaces. The result of the programme was the MiG-25 interceptor, and this was fully developed and placed in service despite the cancellation of the B-70 bomber programme in 1963.

The MiG bureau's only other completely new fighter so far has been the MiG-29, and although this was undoubtedly preceeded by prototypes, these remain unknown in the West.

During World War II the Sukhoi bureau had seemed always to be the bridesmaid rather than the bride, for though it developed several useful types none of these was accepted for large-scale production. The tendency also continued in the years immediately after World War II. The bureau's first turbojet-engined aeroplane was the Su-9 fighter-bomber prototype. This was remarkably similar in overall design to the Messerschmitt Me 262 apart from having a fuselage of vertical oval rather than triangular section, and was powered by a pair of 1984-lb (900-kg) thrust RD-10s. Design began in 1944, and the first prototype flew

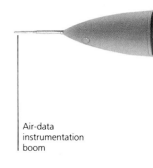

Air-data instrumentation boom

Grouped RWR/ECM/IFF antennae

Outward canted vertical tails

Mikoyan-Gurevich MiG-29
Known in the NATO system of reporting names as the 'Fulcrum', the MiG-29 is one of the most important warplanes in the Soviet inventory of tactical aircraft. The fighter reveals strong evidence of influence by the US-led art of blended aerodynamics in the wide fuselage and leading-edge root extensions, and despite its use of a manual but powered control system, the fighter is extremely agile. Located in front of the windscreen is an infra-red search and track sensor that provides the pilot with a long-range target-acquistion capability even when his main radar is inoperative or jammed.

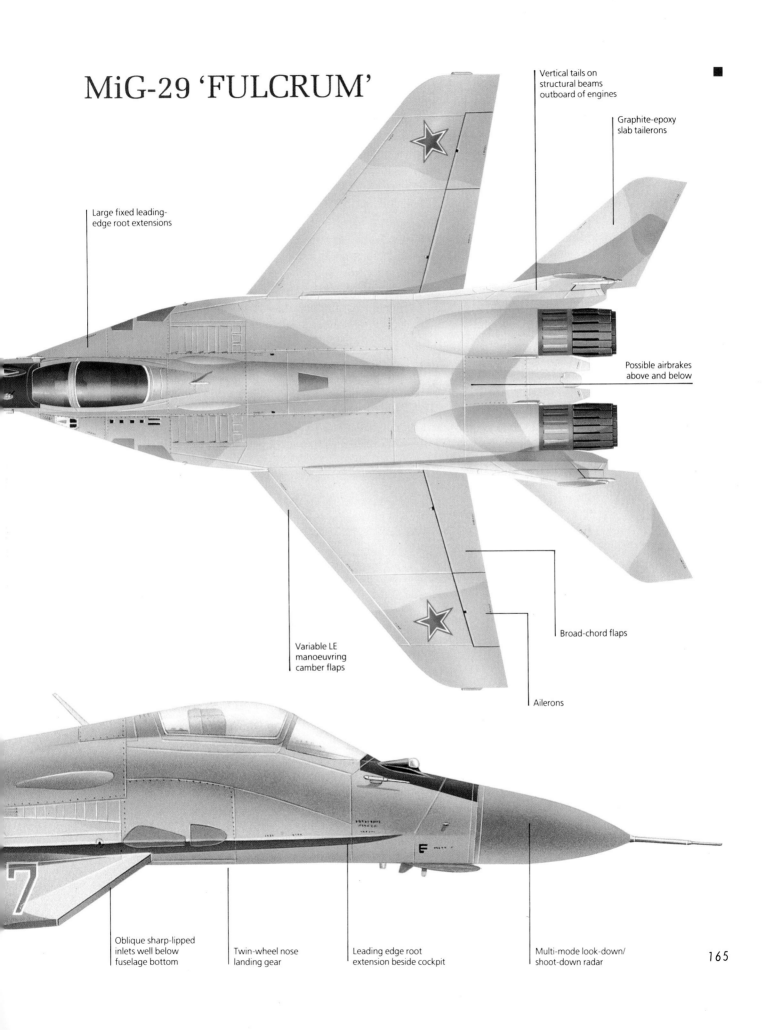

in the summer of 1946. Performance and flight characteristics were impressive, and the type was recommended for production: unfortunately for the bureau there was no available production capability.

The following Su-11 was remarkably similar apart from its use of 2866-lb (1300-kg) thrust Lyul'ka TR-1 turbojets, a slightly larger wing and a pressurized cockpit. Flight trials were started in October 1947, but no production was planned because of inherent problems with the TR-1. Further development of the basic design concept followed with the Su-13, which introduced a slightly swept tailplane and 3505-lb (1590-kg) thrust RD-500 engines. It soon became clear that the design would offer no significant advantages over the Su-11, and not even a prototype was completed.

In January 1948 the Soviet air force issued a requirement for an all-weather interceptor, and among the several designs submitted was one from Sukhoi.

The bureau's next design was the Aircraft R prototype for a planned Su-17 supersonic fighter. But in November 1949 the Sukhoi bureau was closed down by Stalin and the almost complete prototype was cancelled. Sukhoi and most of his team were transferred to the Tupolev bureau and continued work in the development of the aerodynamic and structural features required for supersonic fighters. In 1953 Stalin died, and Sukhoi's request for his own design bureau was then granted: this produced a new sequence of numerical designators that leads to considerable confusion in the identification of Sukhoi aircraft. The first result of the bureau's re-established independence was a series of swept-wing and tailed delta prototypes in the S and T series respectively that paralleled the MiG bureau's Ye-2 and Ye-5 prototypes with either the 58° to 62° swept wing, or the 57° or 58° tailed delta configurations evolved by the Central Aerodynamics and Hydrodynamics Institute.

BELOW. The Sukhoi Su-15 was designed to meet the same all-weather interceptor specification as the Lavochkin La-200 and Mikoyan-Gurevich I-320 and used two turbojets in a tandem arrangement, with one exhaust in line with the wing roots and the other at the tail.

This was developed as the Aircraft P prototype with 35° swept flying surfaces and the same tandem arrangement of engines as used in the competing La-200 and I-320. Sukhoi was the first of the competing bureaux to get a prototype into the air, in January 1949. The prototype revealed good performance, but clearly suffered from problems and when the aeroplane was lost as a result of flutter-induced structural failure the programme for a possible Su-15 production model was cancelled.

The S-1 was an aerodynamically clean type with a low/mid-set wing swept at 62° and a 14,330-lb (6500-kg) thrust Lyul'ka AL-7 turbojet aspirated via a circular nose inlet with a conical inlet centrebody that translated in and out to regulate the supersonic airflow through the inlet; this was the first such installation in a Soviet aeroplane. The type was similar in overall terms to the I-380 from the MiG bureau, and first flew in late 1955. Though short on range because of the thirst of its engine, the S-1 proved to have

ABOVE AND RIGHT. The Sukhoi Su-25 is a close support aeroplane of firmly subsonic performance, and its overall design was undoubtedly influenced by that of the Northrop YA-9, the prototype that lost to the Fairchild Republic YA-10 in the US Air Force's earlier competition for a tank-killing and battlefield close support aeroplane.

admirable performance and handling characteristics. The production offspring of this prototype was the Su-7 fighter-bomber.

In parallel with the S-1, the T-3 was developed with the tailed delta configuration that the bureau thought better suited to the interceptor role than the swept configuration, which was thought to offer its optimum capabilities in the ground-attack role. Very considerable effort was made to produce an inlet best suited to the supersonic regime, and no fewer than 12 inlet arrangements were designed and flown in various prototypes. The T-3 first flew early in 1956, and comparative trials against the S-1 confirmed MiG's experience with the Ye-2 and Ye-5: the tailed delta offered advantages in performance at high altitudes, whereas the swept configuration had both performance and handling superiorities at lower altitudes. Various other prototypes such as the PT-9 validated the chosen inlet system in a practical form,

paving the way for the Su-9 and Su-11 interceptors.

The bureau was also attracted by the possible advantages of lateral inlets, which would leave the nose free for search radar, and developed the T-49 prototype with such a forward fuselage. This confirmed the practicality of the concept, and further evolution produced

SU-25 'FROGFOOT'

Known in the NATO system of reporting names as the 'Frogfoot', the Su-25 is the Soviets' most modern battlefield close support aeroplane. The type was clearly influenced in design by the Northrop YA-9, losing contender in the US Air Force's competition for a battlefield close support and tank-killer aeroplane, and is optimized for considerable agility and survivability at low altitude while carrying a heavy weapon load. High performance is not vital in this role, and the Su-25 is firmly subsonic.

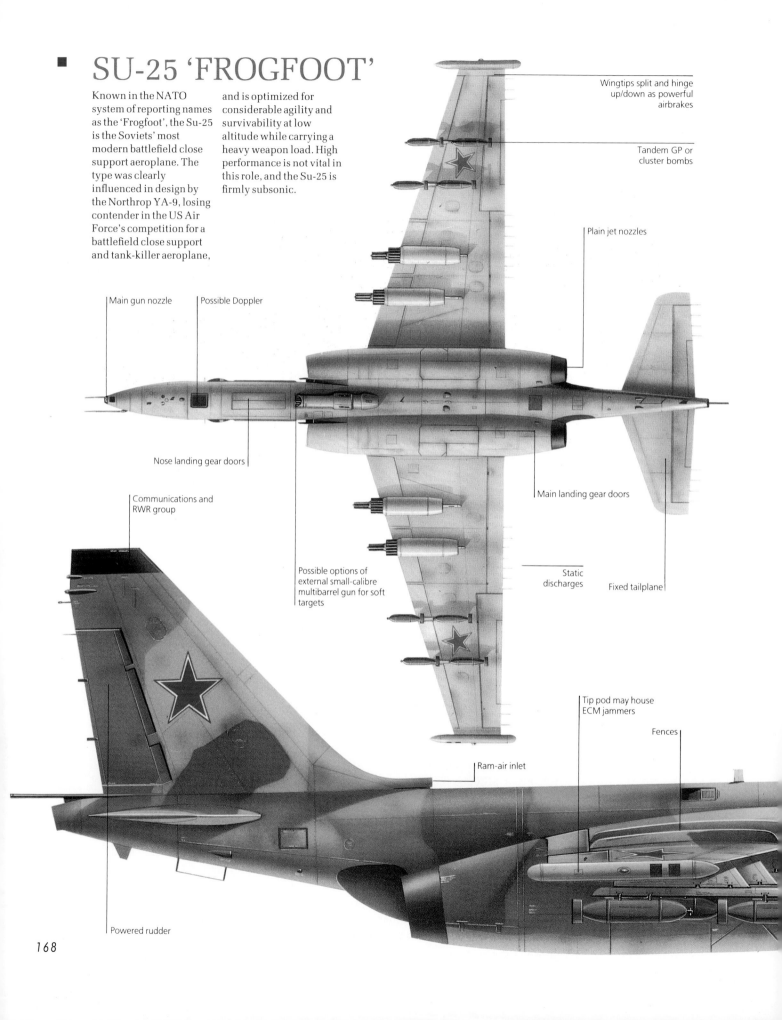

the P-1 prototype for an interceptor for the collision-course intercept role much favoured by the Soviet air force, which believed in tight control of its interceptors. The P-1 flew in 1957, but for unknown reasons did not pass the prototype stage. At much the same time the bureau produced its T-37 prototype meeting the requirement filled by the Ye-266. The T-37 was flown in 1960, but was not designed as anything but a research type.

Features of the T-37 are apparent in Sukhoi's next fighter, the Su-15. This was produced to meet a requirement for an interceptor with Mach 2.5 performance and capability for automatic direction onto the target, and was ultimately developed into the Su-21 that is still an important Soviet air force asset.

In its Su-15/21 programme the Sukhoi bureau decided not to emulate MiG in considering variable geometry for such a fighter, but instead conceived the type as operating from the many long runways available in the USSR. The Sukhoi bureau did not ignore the advantages of variable geometry for improving field performance, however. Instead it opted to consider such a planform for an evolutionary development of the classic Su-7 ground-attack fighter, whose poor payload/range performance could perhaps be transformed by a limited form of variable geometry. It was clear that provision of full variable-geometry wings would require a structural redesign of the fuselage as well as the wings, and was thus impractical. The bureau therefore selected a partial variable-geometry layout in which only the outer wings were pivoted, and this arrangement was used on the S-221 prototype, which was evaluated as the Su-7IG. The modification radically improved the type's payload/range equation, and the type entered production as a type known variously as the Su-17, Su-20 and Su-22 according to model and engine.

Sukhoi has also produced the Su-25 close support and Su-27 fighter aircraft, but as yet few details have been revealed

ABOVE. The portly fuselage of the Yakovlev Yak-19 fighter suggests that the type was powered by a centrifugal-flow turbojet, but the evidence suggests that only an axial-flow type was used. The type was not even considered for production as its engine, derived from a German unit of World War II, was considered obsolete by 1947.

RIGHT. The Yakovlev Yak-30 was designed to meet a 1946 requirement for an interceptor able to operate from rough airfields yet still reach high subsonic speed. The prototype proved to have good handling and adequate performance, but was considered slightly inferior to the Mikoyan-Gurevich MiG-15.

about the prototype development of these important Soviet tactical aircraft.

The fourth Soviet organization involved in the design of fighters is the Yaklovlev bureau, which was responsible for the USSR's first turbojet-powered aeroplane. This was the Yak-15, a comparatively straightforward evolution of the Yak-3 piston-engined fighter with a pod-and-boom fuselage whose deep forward pod accommodated a 1984-lb (900-kg) thrust RD-10 turbojet, the Soviet version of the Junkers Jumo 004B. The first prototype ws ready in October 1945, but the first flight was delayed to April 1946, when the toss of a coin decided that of the two available aircraft the MiG-9 prototype should fly first. Some 280 production aircraft followed, and this type was instrumental in paving the way of turbojet power into Soviet service. The Yak-17 was in essence a product-improved Yak-15 with features such as tricycle rather than tailwheel landing gear. The prototype flew in early 1947, and 430 production aircraft followed.

The bureau's first design intended from the outset for turbojet power was the Yak-19, which was also the first Yakovlev design with stressed-skin construction. The design was tailored round the 2425-lb (1100-kg) afterburning thrust RD-10F located in a straight-through design with a nose inlet and tail exhaust, but was completely lacking in any flair. The prototype first flew in early 1947, but was not even offered for official testing as the bureau appreciated the obsolescence of the fighter's overall concept.

Greater things were expected from the Yak-23, whose design reverted to the underslung turbojet, in this instance the 3505-lb (1590-kg) thrust RD-500 based on a British engine, the centrifugal-flow Rolls-Royce Derwent V. The airframe and aerodynamics were conventional, and the first prototype was flown in June 1947. Large orders were placed, but ony 310 aircraft were completed after it became clear that better overall capabilities were being offered by the conceptually more advanced MiG-15.

The Yak-25 was evolved from the Yak-19 with the RD-500 engine and a swept tail amongst other modifications. The prototype first flew in October 1947 and

BELOW. After extensive trials with the Yak-36 prototype series, the Yakovlev Yak-38 was evolved as a STOVL combat type with two forward-mounted lift jets and one aft-mounted thrust-vectoring jet. The Yak-38 provided the Soviet navy with its first carrierborne fixed-wing aircraft.

RIGHT. Despite the fact that its number in the MiG sequence is higher than that of the MiG-29, suggesting a more modern design, the Mikoyan-Gurevich MiG-31 is in fact an updated version of the MiG-25. The MiG-25 was conceived as an interceptor able to catch the Americans' planned North American B-70 Valkyrie Mach 3 strategic bomber, which was cancelled. Despite the apparent absence of targets flying at very high speeds at very high altitudes, the MiG-25 reached production status as a fighter and reconnaissance aeroplane. The MiG-31 version has lower flight performance, but carries the radar and weapons for effective acquisition and engagement of targets flying lower than itself.

proved exceptional in performance and agility, the former despite the type's straight wings. Even so, the fact that the MiG-15 had been ordered in large numbers meant that there was no service requirement for the Yak-25, excellent though it was. The Yak-25's basic design was then revised to incorporate 35° swept wings, and this Yak-30 prototype first flew in September 1948, again falling foul of faith vested in the MiG-15 and failing to secure a production order. In an almost despairing effort to overmatch the MiG-15, the Yakovlev bureau completely overhauled the basic design to produce the Yak-50 with a 5952-lb (2700-kg) thrust Klimov VK-1 turbojet, flying surfaces swept at 45°, and tandem landing gear under the fuselage with two stabilizing outriggers under the wingtips. The prototype flew in July 1949 and proved itself to possess superlative performance in speed and rate of climb, as well as phenomenal agility. Yet again, however, the industrial commitment to the MiG-15 proved decisive, and Yakovlev's latest fighter failed to find a production slot.

The bureau next turned its attentions to supersonic flight and produced the Yak-1000 prototype for an outright interceptor. The arrangement of tandem main landing wheel units and outrigger stabilizers was again used and the pilot was accommodated on a semi-reclining ejector seat to reduce frontal area. The fuselage was tailored round the 11,023-lb (5000-kg) thrust Lyul'ka AL-5 axial-flow turbojet in a straight-through installation, and the highly swept flying surfaces included a mid-set cropped delta wing and an empennage with the tailplane set about two-thirds of the way up the vertical surface. the Yak-1000 was taxied in 1950, but is believed never to have flown. It is thought that very high levels of instability were suspected, and the programme was cancelled.

The Yakovlev bureau finally hit on the right combination of factors in the Yak-25. In response to the November 1951 all-weather interceptor requirement that also produced the La-200 and I-320, Yakovlev produced a scaled-up version of the Yak-50 with its single fuselage-mounted

RIGHT. The Mikoyan-Gurevich MiG-29 is one of the USSR's two latest fighters, and though it retains the mechanical control system of the previous generation of fighters, its blended contours and advanced aerodynamics bespeak a considerable quantity of research and development in wind tunnels and on prototype aircraft.

turbojet replaced by two 4850-lb (2200-kg) thrust axial-flow turbojets mounted in wing nacelles so that the nose could be left free for the massive radar demanded by the requirement. The prototype had flying surfaces swept at 45° and first flew in 1953. The trials were very successful, and a production programme took the type right through toward the end of the 1960s in role-differentiated models designated Yak-25, Yak-26 and Yak-27 with swept wings and on the Yak-25RD high-altitude reconnaisance version with straight wings. Further evolution of the same basic design with a much refined aerodynamic and structural design produced the supersonic Yak-28 produced in tactical strike, interceptor and electronic escort models.

In 1962 the Yakovlev bureau was chosen to produce the first Soviet vertical take-off and landing aeroplane. Initial consideration was given to a composite arrangement of lift jets and a cruise engine, but it was finally decided to use two Tumanskii turbojets with vectoring nozzles on the centre of gravity to provide direct lift or forward thrust as required. The airframe designed for the new Yak-36 was necessarily broad to accommodate the side-by-side engines, used the now-standard arrangement of tandem main units on the centreline together with stabilizing outriggers at the wingtips, and was completely conventional as only high subsonic speeds were envisaged. Unlike the British Hawker P.1127 type, which has four vectoring nozzles fed from a single turbofan, the Soviet arrangement of just two nozzles each directly the thrust of its own engine was potentially dangerous, for the failure of either engine in the lift mode would have resulted in an uncontrollable roll. Hovering control was provided by reaction jets in the wingtip pods, the tail and the long nose boom. The type first flew in the mid-1960s, and trials with at least 12 such prototypes paved the way for the Yak-38 VTOL naval aeroplane, which has a composite powerplant with one vectored thrust turbojet in the rear fuselage, and two lift turbojets in the forward fuselage.

Further information about Soviet fighter prototypes is almost completely lacking, but it is certain that further development is being undertaken to produce a generation of aircraft to supersede the MiG-29 and Su-27 as the USSR's primary fighters to match the new generation of machines being developed in the West. Notable throughout the process of Soviet fighter design has been the steady way in which the USSR has closed the technological leeway between its fighters and those of the West. By the time of the MiG-29's and Su-27's debut in the mid-1980s the West's technological advantage had all but disappeared (indeed had been matched in some areas), and it is possible that the next generation will see the USSR match the West or even push slightly ahead of it.

BELOW. The Sukhoi Su-27 is the MiG-29's companion as the USSR's most important fighter assets. It is a larger machine than the MiG-29 but while again using blended aerodynamics and an advanced structure, has the advantage of an electronic 'fly-by-wire' control system.

FIGHTERS FROM OTHER COUNTRIES

ABOVE. The Hawker P.1040 was one of several British fighter designs at the end of World War II that failed to progress past the prototype stage.

UK

The only other countries to have produced advanced turbine-powered fighter prototypes in the period up to 1980 have been several European powers and to a lesser extent Canada. The 10 years after World War II have rightly been described as years of lost opportunity for the UK. The country emerged from the war as the leading Allied exponent of turbojet power, and the only one of the three principal Allies to have fielded an operational turbojet-powered fighter. This Gloster Meteor was kept in development after the war, and soon supplemented by the lighter but more agile de Havilland Vampire. However, the possibilities for the future were then squandered by the country's moral exhaustion, which combined with financial difficulties to produce a political climate in which far-sighted projects were inevitably whittled back to limited research objectives and then, as often as not, cancelled.

Typical of the concepts caught in this process was the Miles M.52 supersonic fighter. This was conceived as early as 1943 in a specification calling for an interceptor capable of 1000 mph (1609 km/h) at 36,000 ft (10,975 m). This was a remarkably bold concept, in effect leapfrogging the transonic stage so that the Royal Air Force might be able to proceed from firmly subsonic to firmly supersonic fighters in a single step. The M.52 was based on a cylindrical fuselage in whose forward opening the cockpit was located as a centrebody, thereby creating an annular inlet; the cockpit was designed for separation from the fuselage by explosive charges in an emergency, the pilot then baling out under safe conditions. The 2000-lb (907-kg) thrust Power Jets (Whittle) W.2/700 turbojet was located in the central fuselage, boosted by the later addition of a rear fan (effectively turning the engine into a turbofan) and combustion cans in the rear of the duct (effectively turning the engine into an afterburning type). A straight but very thin wing was located in the mid-set position, and used a special aerofoil of bi-convex section; and an all-moving tailplane was planned. Construction of the first of three prototypes was imminent when, in February 1946, the complete project was cancelled as the British government had decided that it would be safer as well as more economical to proceed toward supersonic flight through wing-tunnel experimentation rather than manned aircraft.

Gloster was also concerned with the development of more advanced turbojet-powered fighters, but its next two aircraft were abortive: the G.42 (more frequently

ABOVE. With modestly swept wings the P.1040 became the Hawker P.1052, but again this failed to move from the prototype to the production stage.

known as the E.1/44 after its specification), was a small fighter powered by a single Rolls-Royce Nene turbojet aspirated via lateral inlets. The G.42 first flew in 1947 but produced no production orders. The CXP-1001 resulted from Chinese Nationalist ambitions, but had got no further than the mock-up stage when the design's sponsors lost the Chinese Civil War in 1949 and dropped the project. Greater success attended the company's next design, the GA.5 that was accepted for RAF service as the Javelin all-weather interceptor.

Hawker was one of the major British producers of fighters in World War II, and by 1943 was actively involved in the design of turbojet-powered successors to the Tempest and Fury/Sea Fury piston-engined fighters. The company's faith was pinned mostly on Rolls-Royce engines, notably the B.40 and B.41, and early projects were the P.1031 night fighter with a single B.40 and the P.1035 version of the Fury with a B.41. There were also three bomber projects.

The most promising of these projects was the P.1035, which was gradually evolved away from its Fury origins with a long jetpipe exhausting under the tail to become the P.1040 with a bifurcated system so that the engine could exhaust through the fuselage sides just aft of the wing via two short jetpipes. At this stage the RAF lost interest in the type, which was thought to offer inadequate performance advantages over the Meteor F.Mk 4, but it was saved by Royal Navy interest, which resulted in the P.1040's development as the Sea Hawk carrierborne fighter and fighter-bomber.

From the P.1040's basic concept, a string of other types evolved. First came the P.1047, a scheme for the P.1040 revised with wings swept at 35° (though the tail surfaces remained straight) and powered by a rocket rather than a turbojet, but from this somewhat impractical scheme emerged the P.1052 prototype that first flew in November 1948 with a 5000-lb (2268-kg) thrust Rolls-Royce Nene turbojet. No production order was forthcoming, but Hawker still thought that the type had merit and, in response to an Australian requirement, further developed the type as the P.1081 with swept tail surfaces and a straight-through engine installation using a jetpipe adapted from that of the

Supermarine Attacker naval fighter. The prototype first flew in June 1950 with a Nene turbojet, though the 6250-lb (2835-kg) thrust Rolls-Royce Tay turbojet was planned for any production variant. The sole prototype crashed in April 1951, however, and the programme was terminated.

The British had not lost interest in rocket power during the period, though emphasis was now placed on a composite powerplant using a turbojet for endurance and a liquid-propellant rocket for maximum climb rate and speed in the interception role. Thus the P.1047 concept was revived as the P.1072, still with straight flying surfaces but now powered by a 5000-lb (2268-kg) thrust Nene turbojet exhausting via bifurcated ducts in the wing roots and supplemented by a 2000/lb (907-kg) thrust Armstrong Siddeley Snarler rocket exhausting under the tail. After a conventional first flight in November 1950, the rocket was successfully used four days later. By this time the British were coming to favour afterburning turbojets over composite powerplants, and only a few more test flights were undertaken. Even so, the development of these prototypes had given Hawker invaluable experience into the aerodynamics and structural features required for transonic fighters, and the company's next design was the P.1067 prototype for the superlative Hunter, which was the arguably the finest transonic fighter ever produced. Plans were also laid for a supersonic development of the Hunter as the P.1083 with an afterburning version of the Rolls-Royce Avon turbojet. The prototype was 80 per cent complete when the project was cancelled.

The two other British producers of fighters in this period were de Havilland and Supermarine. De Havilland developed several important types such as the Venom and Sea Vixen based aerodynamically on the concept pioneered in the Vampire: a central nacelle and a tail unit supported on booms reaching back from the wings. There were problems with these designs, but both moved forward to the production stage without undue difficulty.

Supermarine had a more problematical time. The starting point of the company's vestures into turbojet-powered warplanes

BELOW. When fitted with an Armstrong Siddeley Snarler liquid-propellant booster rocket in the tail, the P.1040 became the Hawker P.1072 and first flew in November 1950. Performance was improved considerably, but it was appreciated that liquid-propellant rockets were subject to a number of technical limitations and the programme was soon closed down.

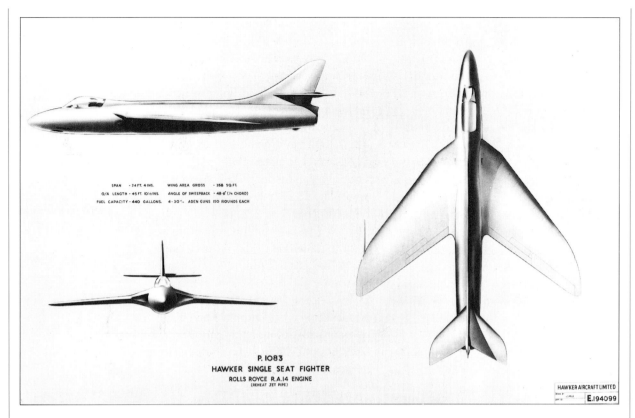

began with the Attacker, which became the first such aeroplane adopted by the Royal Navy. The Attacker exemplified the notion of the piston-engined aeroplane 'translated' into turbo-powered form, and as such ws elarly only an interim type. From the Attacker, however, Supermarne developed the Type 510, which was the Attacker fitted with swept flying surfaces for a first flight in December 1948, and then the Type 535, which was the Type 510 with a longer nose and tricucle rather than tailwheel landing gear. The Type 535 first flew in August 1950, and proved sufficiently impressive for a 100-aircraft order to be planned if the same company's Type 541 proved a failure. The Type 541 was in fact the prototype for the Swift production model, and flew in August 1951. The type was ordered into production without delay as the Korean War was proving the obsolescence of current RAF fighters, and the Swift began to enter service in February 1954 as the RAF's first swept-wing fighter. Further evolution of the Swift was planned as the Type 545 to provide supersonic performance. It was the potential of this project that led to the cancellation of the Hawker P.1083, despite the fact that the Supermarine design was a higher-risk development that promised performance inferior to that of the P.1083. By comparison with the Swift, the Type 545 featured an area-ruled fuselage and a crescent-shaped wing whose leading-edge sweep declined from 50° at the roots to 40° at mid-span and 30° on the outer panels. So many problems were encountered with the Swift in operational service, however, that in 1955 all work on the Type 545 was cancelled.

This left the UK without a supersonic fighter in development, and the RAF then turned hastily to the English Electric P.1 supersonic research aeroplane, which in its P.1B form was turned into the prototype for the short-ranged but otherwise impressive Lightning interceptor. The P.1B first flew in April 1957, and after protracted development the Lightning finally entered service in 1960.

The UK thus moved straight from the subsonic to the truly supersonic fighter without the intermediate step of a

ABOVE. As its lines indicate, the Hawker P.1083 would have been a mildly supersonic successor to the classic Hunter and as such would probably have capitalized on the Hunter's very considerable commercial success. However, with the airframe of the first prototype 80 per cent complete the project was cancelled by the British government.

transonic type. This was a considerable technical achievement, but ultimately it proved self-defeating in economic terms: the huge export success of the Hunter could not be exploited when the type's operators demanded a transonic successor such as the cancelled P.1083, and this led to a weakening of the British aero industry in overall terms. At the time this tendency was hardly discernible, for the British were in the throes of a major effort to produce fighters capable of dealing with the perceived threat of Soviet manned bombers in the early 1960s. These were envisaged in two forms as a Mach 0.9 heavy bomber and a Mach 2 medium bomber, each operating at an altitude of more than 60,000 ft (18,290 m).

The RAF's resulting requirement for an interceptor with extraordinarily climb rate caused designers to turn their attention once more to the rocket motor, of which two types had been under development since 1946. The Armstrong Siddeley Snarler was a 2000-lb (907-kg) thrust type running on methyl alcohol, water and liquid oxygen, while the de havilland Sprite was a 5000-lb (2268-kg) thrust unit running on high-test hydrogen peroxide. The Snarler was tested in the Hawker P.1040 prototype that became the P.1072, but though climb performance was transmogrified the Snarler had technical problems and was cancelled. The Sprite was test flown under the de Havilland Comet airliner prototype from April 1951.

From the Snarler was evolved the Screamer, the UK's first variable-thrust rocket, while from the Sprite was developed another second-generation rocket, the Spectre. In 1952 the specification for a rocket-powered interceptor was issued, its requirements including high level speed, very rapid climb after a short take-off, and an armament of four air-to-air missiles. Several companies responded, the two most promising designs being the Avro Type 720 and the Saunders-Roe SR.53. Each used a composite powerplant with a turbojet for endurance and a liquid-propellant rocket for maximum speed and climb performance. Prototypes of each type were ordered, the Type 720 with a powerplant of one Armstrong Siddeley Viper turbojet and one Screamer rocket motor, the latter running on kerosene and liquid oxygen, and the SR.53 with the Viper complemented by a Spectre rocket motor. Both machines underwent considerable development before the final design was approached, the definitive Type 728 and P.177 being planned round a combination of the 8000-lb (3629-kg) thrust de Havilland PS.38 (later Gyron Junior) turbojet and the identically rated Spectre rocket motor. Ultimately only the SR.53 flew, almost three years later than had been optimistically planned, in July 1957. The SR.53 revealed hugely impressive climb performance, but the project was cancelled after the unexplained loss of second prototype in March 1958. The much improved P.177 version had already succumbed to the widespread cuts that followed the notorious 1957 British decision that the future of aerial war lay with missles rather than manned aircraft.

Other British warplanes of the period that remained purely paper exercises were the Fairey F.155T Mach 2.5 manned interceptor, the Hawker P.1121 Mach 2.25 strike and air-superiority fighter, and the Hawker P.1129 Mach 2.3 tactical strike and reconnaissance aeroplane. Greater success in the short term, attended the incredibly advanced English Electric TSR-2 Mach 2 tactical strike and reconnaissance aeroplane, evolved from the P.17A concept. The first TSR-2 prototype flew in September 1964, and although the programme was faced by considerable technical difficulties these were well on the way to solution when the programme was cancelled in April 1965.

FRANCE

At the end of World War II France was faced by the immense task of rebuilding her armed forces and revitalizing her national defence capability. The task was undertaken with great vigour, and the French aero industry was soon suggesting a number of fighter types to follow the turbojet-powered research aircraft it was producing to validate a host of technologies new to the French. The first pure-jet fighter developed in France was the Sud-Ouest SO.6025 Espadon, a straight-winged interim type that made its initial flight during November 1948 in the form of the SO.6020 whose Rolls-Royce Nene 102 turbojet was aspirated via a ventral inlet under the rear fuselage. This proved completely unsatisfactory, and the second SO.6020 featured flush inlets on the fuselage sides aft of the wing trailing edge. The third SO.6020, shortly redesignated SO.6025, had a long ventral inlet in whose rear portion was accommodated a SEPR 251 liquid-propellant booster rocket. Considerably more development followed, but by 1953 the type was clearly obsolescent and further work was abandoned.

Another French prototype of the period was the Sud-Est SE.2410 Grognard attack fighter, an interesting type whose two Nene 101 turbojets were aspirated via a dorsal inlet. The Grognard I first prototype flew in April 1950 as a single-seater with wings swept at 47°, while the Grognard II second prototype was a two-seater with wings swept at 32°. Both designs suffered from flutter-induced problems, and was cancelled when the Sud-Ouest Vautour offered considerably greater capabilities. The Aerocentre NC.1071 attack aeroplane prototype, France's initial twin-jet design, first flew in October 1948. The long nacelles of the Nene turbojets supported at their rear the twin vertical surfaces that were joined at their upper ends by a tailplane which was thus kept well above the aerodynamic disturbance of the bulky central nacelle. Aerocentre went into liquidation during 1949 and development of the NC.1701 was ended. Another Aerocentre design, the NC.1080 naval fighter, became the Nord 2200 that first flew in December 1949 but was discontinued in 1952.

Another naval fighter prototype was the Arsenal VG.90, which was developed from the VG.70 experimental aeroplane powered by a Junkers Jumo 004 turbojet. The first two VG.90 prototypes were built of wood, and the earlier of these made its maiden flight in September 1949: both aircraft were lost in air accidents, and the all-metal third prototype never flew.

Greater success attended the efforts of Dassault, whose Ouragan was France's first operational jet fighter. From the straight-winged Ouragan the company evolved the swept-wing Mystere I that first flew in February 1951 with a Nene 104B turbojet and was followed by two

The Sud-Ouest SO.6025 Espadon was France's first jet fighter, and an unusual appearance was given to the aeroplane's underside by the location, at the rear of the ventral air inlet, of a SEPR 251 liquid-propellant booster rocket.

Mystere IIA prototypes each powered by a Rolls-Royce Tay turbojet. These paved the way for the Mystere IIC production model powered by the SCECMA Atar 101 turbojet. The Mystere III was a two-seat all-weather development with lateral inlets to leave the nose free for radar, and the sole prototype of this model flew in July 1953. In parallel with this development stream the company had been working on a more advanced interceptor, the Mystere IV. This overtook the structurally troubled Mystere II on the production line and paved the way for the Super Mystere B1 that in March 1955 became the first production-standard European fighter to achieve more than Mach 1 in level flight.

These were all conventionally configured aircraft, but in the MD.550 Mirage I the company adopted the layout for which it is now best known, namely the tailless delta. The Mirage I was designed to meet a requirement for a lightweight interceptor and first in June 1955 with two Armstrong Siddeley Viper turbojets, subsequently supplemented by a SEPR 66 liquid-propellant booster rocket to produce a level speed of mach 1.3. Further development was required to produce an effective machine, and after the Mirage II with twin Turbomeca Gabizo turbojets had been abandoned, the company produced the classic Mirage III supersonic fighter.

Nord was another protagonist of the tailless delta configuration, and in August 1954 is N.1402 Gerfaut I prototype with the Atar 101D3 turbojet became the first aeroplane to exceed Mach 1 in level flight without an afterburner or a supplementary rocket motor. Gerfaut IB

BELOW. The product of a protracted development programme, the Dassault Mystere IVA was a capable though unexceptional fighter-bomber that was also used for a number of development tasks.

and Gerfaut II prototypes were also produced, and these paved the way for the N.1500 Griffon prototype of interceptor potential. This flew in September 1955 with an afterburning Atar 101F turbojet, but was soon fitted with the planned combination powerplant comprising an Atar 101E turbojet and a Nord ramjet to produce the Griffon II. This was capable of supersonic performance on its ramjet, but was not developed into an operational type.

Another French prototype with combination power (in this instance turbojet and rocket) was the Sud-Ouest SO.9000 Trident. Construction of the Trident I prototype began in 1951, and a time when the vogue was for swept of delta planforms the design team opted for a straight but very thin wing. The Trident I made its first flight in March 1953 on the power of two Turbomeca Marbore turbojets mounted in wingtip nacelles. The second prototype was lost on its first flight, and the first machine was then re-engined with Viper 5 turbojets so that it would have sufficient power to get into the air with its fuselage-mounted engine, a SEPR 481 triple-chamber rocket located in the tail. This model flew in May 1955, and further development was entrusted to Dassault, which built two SO.9050 Trident IIs as the prototypes for a planned operational version. This Trident II reached Mach 1.7 but both prototypes were lost in aerial accidents, the second confirming the inherent dangers of a rocket engine with its immensely volatile fuel.

This prolific period of French aircraft design also saw the development of the small delta-winged Sud-Est SE.212 Durandal as a possible interceptor, and the Sud-Est SE.5000 Bardoudeur as a fighter-bomber. The latter was designed for deployment close to the front line, and featured a landing gear arrangement of twin skids under the fuselage, which dictated the use of a wheeled but jettisonable trolley for take-off.

In common with other countries, France became increasingly concerned during the 1960s with developing warplanes that not only offered superior maintainability and operational flexibility, but were also less reliant on the long paved runways that were obviously vulnerable to conventional as well as nuclear attack. The two approaches that offered the best chances of success were the variable-geometry wing planform (with its other benefit of enhanced payload/range performance), and the VTOL concept.

ABOVE. The Sud-Ouest SO.9000 Trident was a notably ambitious fighter prototype whose mixed powerplant comprised a liquid-propellant rocket in the tail and two small turbojets in the wingtip pods.

In 1965, the UK and France agreed to the collaborative development of a supersonic aeroplane that could be used in the attack and a trainer roles (the SEPECAT Jaguar), and in the same year it was also decided to proceed with the development of the Anglo-French Variable-Geometry multi-role warplane. This was never a fully viable project, and in 1967 the French withdrew. It is interesting to note, however, that many of the features from this ambitious warplane re-emerged later in the decade in the design for the British, Italian and West Germany Multi-Role Combat Aircraft that finally matured as the Panavia Tornado.

Despite the failure of the AFVG project, France was still interested in the variable-geometry concept and pushed ahead with two indigenous types, the Dassault Mirage G and the Mirage G.8, the former flying in November 1967 and the latter in May 1971. Neither entered production.

ABOVE. The first prototype of the Dassault-Breguet Mirage 2000 (left) is seen in flight with the prototype Dassault-Breguet Mirage 4000 during 1979. The Mirage 2000 has gone on to a creditable production career while the Mirage 4000 has remained only a prototype.

LEFT AND RIGHT. The two Dassault Mirage G8 VTOL prototypes reveal the minimum and maximum sweep positions of this French variable-geometry design.

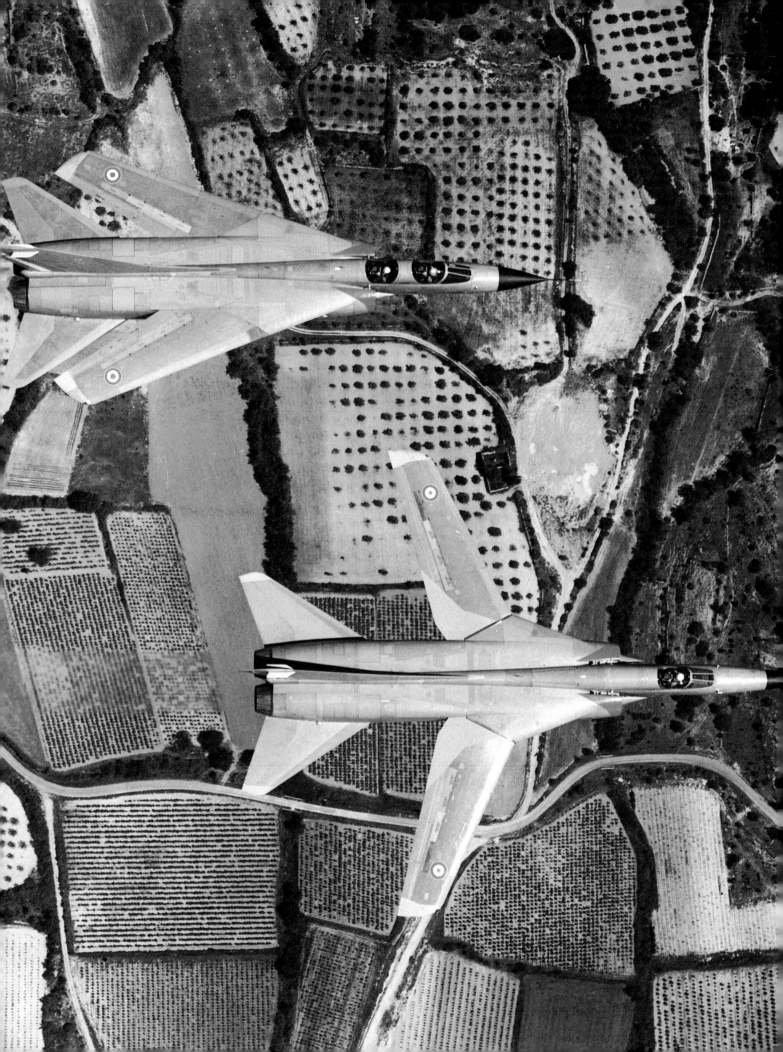

ACX 'RAFALE'

Dassault-Breguet Rafale-A
The Rafale-A is a demonstrator for the new generation of French tactical aircraft, the land-based Rafale-D and the carrierborne Rafale-M, which will be slightly smaller and lighter than the demonstrator. The overall configuration reflects the importance of increased agility and a good field of vision for the pilot. Sophisticated electronics are provided to look after the minute-by-minute chores of flight and presents to the pilot only those factors that are essential for survival and combat.

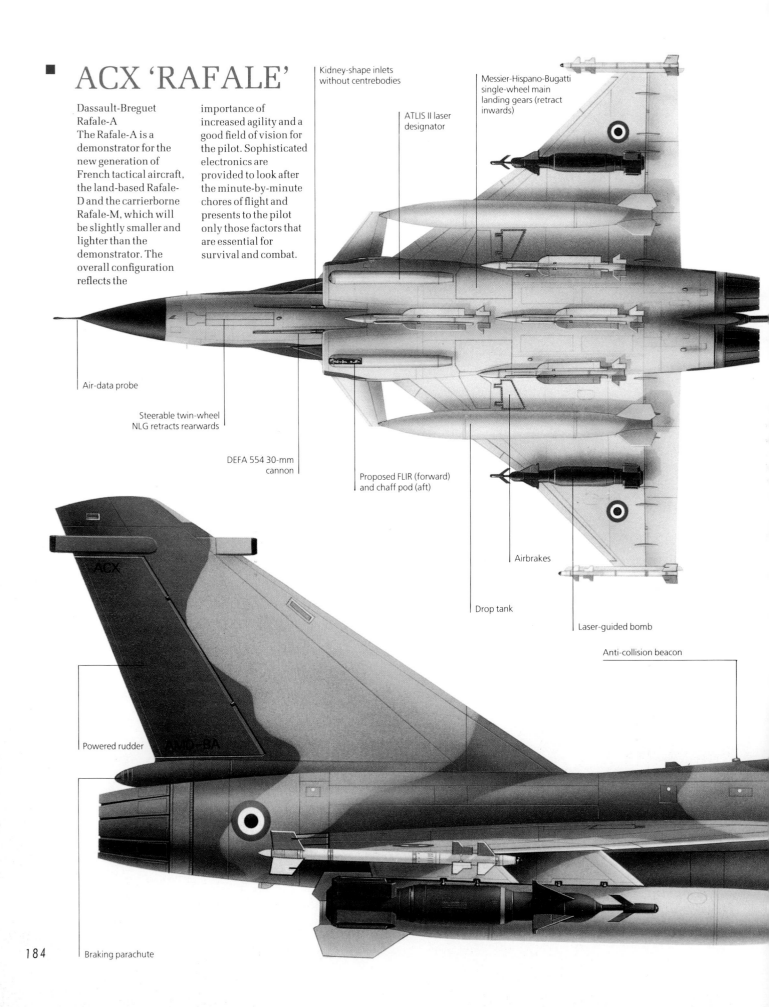

VTOL AIRCRAFT

There was also considerable interest in France and the UK for VTOL aircraft. By the early 1960s considerable experimental work had been undertaken in several countries, most notably the USA where prototypes for a number of VTOL concepts had been flown, but France and the UK went in separate directions at this time.

The French opted for VTOL of the direct-lift type, and produced their first prototype in the form of the Dassault Balzac. This used the cockpit, wings and vertical tail of the Mirage III married to a similar but new fuselage containing a 4850-lb (2200-kg) thrust Bristol Orpheus turbojet for forward propulsion and a battery of no fewer than eight 2160-lb (980-kg) thrust Rolls-Royce RB.108 turbojets mounted vertically for direct lift. These were installed in four groups of two on each side of the centreline fore and aft of the centre of gravity. The Balzac first flew in October 1962 in tethered flight, made its first free flight in the same month and its first transitions in March 1963. The test programme confirmed the viability of the lift/propulsion arrangement, and Dassault moved forward to the Mirage III-V based on the Mirage IIIE strike fighter with its fuselage lengthened to accommodate eight 3525-lb (1600-kg) thrust Rolls-Royce RB.162 lift jets as well as a 13,890-lb (6300-kg) thrust SNECMA TF-104 turbofan for forward propulsion. The Mirage III-V first flew in February 1965, in a tethered flight later followed by a free flight and then a translational flight. The Mirage III-V reached Mach 1.35 with its original propulsion engine, and Mach 2.04 when fitted with the 18,520-lb (8401-kg) thrust General Electric TF30 turbofan. Production was planned, but was finally abandoned in favour of the Mirage F1, a conventional type that had been

BELOW. The Rolls Royce (Bristol Siddeley) Pegasus is the engine used in the revolutionary British Aerospace (Hawker) Harrier STOVL warplane, and this photograph of a test engine shows the starboard rear nozzle, which with its port partner exhausts the hot gasses from the turbine section, in the rear-facing position.

ABOVE. A Hawker Kestrel FGA.Mk 1 evaluation aeroplane shows its port nozzles in the down-facing position.

RIGHT. The Hawker P.1127 prototype for the Kestrel and Harrier series hovers during its initial tethered trials.

produced somewhat ironically to test the armament and propulsion systems for the planned Mirage III-V variant.

The UK's interest in VTOL first took concrete form with the Rolls-Royce Thrust Measuring Rig, otherwise known as 'The Flying Bedstead', and this paved the way for the Short SC.1. This had five RB.108 engines (four of them for direct lift and the fifth for forward propulsion), and first flew in April 1957. Ultimately, however, the British came to the use of vectored thrust for VTOL capability as this removed the need for separate lift engines which added weight and cost while constituting only so much dead

BELOW. The Short SC.1 was the UK's first VTOL aeroplane, a research design whose five Rolls-Royce RB.108 turbojets were located as one propulsion unit and four direct-lift units.

ABOVE. The Dornier Do 31 was an extremely ambitious STOVL transport protrotype in which two Rolls-Royce Pegasus 5-2 thrust-vectoring turbofans were complemented by eight Rolls-Royce RB.162-4 direct-lift turbojets located as four in each wingtip pod.

weight in wingborne flight.

The engine developed for this purpose was the Bristol (now Rolls-Royce) Pegasus, and its first platform was the Hawker P.1127. The airframe was designed round the unique character of its Pegasus engine, which has four nozzles (the forward pair exhausting cold air from the engine's fan stage and the aft pair exhausting hot gases from the engine's turbine stage) arranged two on each side fore and aft of the centre of gravity. The P.1127 first flew in tethered form during October 1960, the first translations being made in September 1961. Such was the promise of the type that a pre-production derivative was built as the Kestrel, and the full production type as the Harrier, which has since matured in somewhat different form as the Anglo-US McDonnell Douglas/British McDonnell Douglas/British Aerospace Harrier II. It had been planned to produce a supersonic production derivative of the P.1127 as the P.1154, but this was cancelled.

ABOVE. The Israel Aircraft Industries Lavi was an extremely sophisticated warplane designed in Israel with a large measure of US financial and technical support, but was cancelled for financial reasons soon after the first prototype had flown.

LEFT. The SNECMA C.450 Coleoptere was an extraordinary VTOL research aeroplane that rose vertically on the power of its SNECMA Atar 101 turbojet before translating into forward flight supported by its annular wing.

Two other VTOL prototypes worth mentioning are a pair of West German aircraft, the Dornier Do 31E and the Entwicklungsring Sud Arbeitsgemeinschaft VJ-101C. The Do 31E was a transport prototype that first flew in February 1967 with two Pegasus vectored-thrust turbofans in underwing nacelles and two removable wingtip pods each containing four 4400-lb (1996-kg) thrust RB.162 turbojets. The VJ-101C was a prototype in fighter configuration that first flew in April 1963 with six 2750-lb (1247-kg) thrust RB.145 turbojets. These were located as two direct-lift units in the fuselage and the other four units in two wingtip nacelles that could be rotated between vertical and horizontal for direct lift and thrust. A second prototype used the 3550-lb (1610-kg) thrust afterburning version of the same engine, and was supersonic in level flight, but as with the Do 31E no production followed. Further West German development resulted in the VFW-Fokker VAK-191B strike and reconnaissance fighter prototype that first flew in September 1971. This was powered by one 10,150-lb (4604-kg) thrust Rolls-Royce/MTU RB.193 vectored-thrust turbojet and two 5577-lb (2530-kg) thrust RB.162 lift turbojets. Despite successful flight trials it was not ordered into production as the Harriers of RAF Germany were already fulfilling the need for which the VAK-191B had been conceived.

There has been something of a lull in VTOL prototype development in recent years, but considerable work has been undertaken on advanced engines for supersonic VTOL aircraft, and it will probably not be long before more advanced aircraft are under prototype evaluation.

LAVI

Israel Aircraft Industries Lavi

Israel has long been concerned about the possibility of interruptions of vital weapon deliveries from overseas suppliers, as a result of political, economic and military action. Israel has thus developed an indigenous aero industry whose most ambitious type to date has been the Lavi, which reached the prototype stage before being cancelled under domestic and US economic pressures. The design was an early example of the modern vogue for the close-coupled canard configuration, and made extensive use of composite materials in its structure. Provision was made for a large and very varied load of weapons, delivered with the aid of a very advanced suite of largely Israeli electronics.

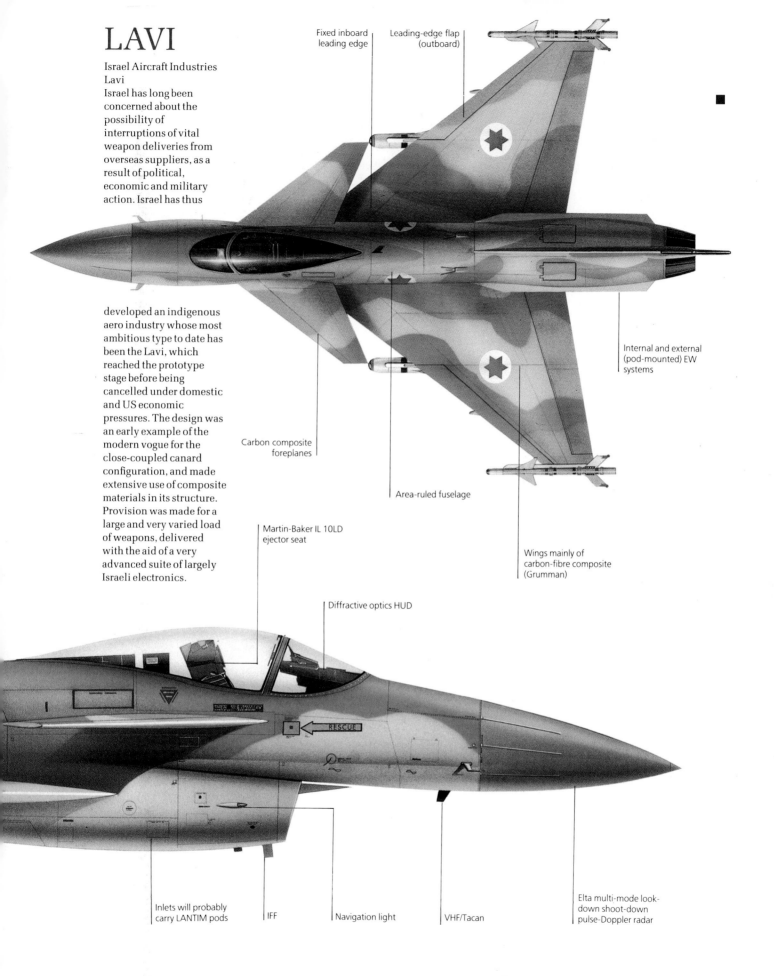

The canard concept has come fully of age in two impressive Swedish warplanes, initially the Saab 37 Viggen and more recently the Saab JAS 39 Gripen illustrated here in artwork form. This offers the same basic combat capabilities as the Vigen in a smaller, lighter, cheaper and considerably more agile airframe.

RESEARCH AIRCRAFT

The previous chapters have been concerned primarily with prototypes that were designed to produce operational aircraft. There remain experimental and research aircraft that have often contributed in major ways to the design and indeed the development of operational aircraft but which were in themselves schemed for investigation of aerodynamic, propulsion or structural features.

USA

The USA has led the way in such aircraft since World War II, for it has had the technical skills, financial strength, interest and far-sightedness to involve itself in basic research of this nature. Within the US aero industry the company that was most deeply concerned with aerodynamic investigation was Bell, now best known for its expertise in the field of light helicopter design but in the period after World War II a major force in the pioneering of high-speed flight.

The Bell X-1 series produced the first flight through the 'sound barrier' when on 14 October 1947 an aeroplane of the type reached Mach 1.015 in the hands of Captain 'Chuck' Yeager. The type was planned specifically for investigation into transonic and supersonic flight, and was based on a cylindrical fuselage that accommodated the 6000-lb (2722-kg) thrust Reaction Motors E6000-C4 rocket

BELOW AND RIGHT. The Bell X-1 series aircraft had straight flying surfaces, but as these were very thin and used on airframes with liquid-propellant rocket propulsion the aircraft were capable of supersonic flight. Indeed, it was in an X-1 that the first supersonic flight was made during October 1947. The X-1A had turbo-driven fuel pumps, and the X-1B was used for research into the thermal characteristics of high-speed flight.

motor and the tanks for its liquid propellants. The flying surfaces were straight but thin, and the first X-1 powered flight was achieved in December 1946 after an air drop from a Boeing B-29 motherplane. The three X-1s were followed by an X-1A with a longer fuselage for greater fuel capacity, a revised cockpit canopy, and turbo pumps replacing the previous pressurized nitrogen fuel feel system: the X-1A reached a speed of Mach 2.435 and an altitude of more than 90,000 ft (27,430 m). The X-1B was used for thermal research, and the X-1D was lost after only one flight. The X-1E had a knife-edge windscreen and wings of 4 per cent rather than 10 per cent thickness/chord ratio.

The Bell X-2 was a development in the field already pioneered by the X-1, with a cylindrical fuselage of K-monel alloy and swept flying surfaces of stainless steel. The powerplant comprised a 15,500-lb (6804-kg) thrust throttlable Curtiss-Wright XLR-25-CW-1 liquid-propellant rocket. Two such aircraft were built, the first being lost in May 1954 when it was jettisoned after an explosion in its Boeing B-50 motherplane. The second aeroplane made the type's first powered flight in November 1955, and recorded an altitude of 126,000 ft (38,465 m) as well as a speed of Mach 3.2, the latter being recorded during the type's fatal last flight in September 1956.

The Bell X-5 was produced to investigate the aerodynamic consequences of altering the wing geometry in flight, and first flew in June 1951. Work on the two X-5s began in 1948, the basis of the design being the Messerschmitt P.1101 prototype that had almost been completed by the Germans at the end of World War II. Powered by a 4900-lb (2222-kg) thrust Allison J35-A-17 turbojet located in the lower fuselage and exhausting under the tail boom, the X-5 had wings that could be varied between in sweep between 20° and 60° in flight, the hydraulic operating mechanism automatically compensating for the

LEFT. The Bell X-5 may be regarded as the world's first aeroplane with a practical variable-geometry wing layout, but was developed as a prototype purely for research purposes.

BELOW. The XF-85 Goblin, which first flew in 1948, was a fighter prototype designed for launch and recovery by the B-36 bomber.

inevitable shift in the position of the design's centre of gravity. Special fairings were also fitted to ensure that the leading and trailing edges of the wing roots presented smooth aerofoil surfaces at all times. The programme yielded much information about the utility of variable-geometry wings for good low-speed handling with the wing in the minimum-sweep position without detriment to high-speed performance with the wings in the maximum-sweep position.

The Bell X-14 was produced as a vertical take-off prototype, and achieved its first hovering flight in February 1957. The airframe was as simple and light as possible, and was characterized by an open cockpit and fixed tailwheel landing gear. In its original form the aeroplane was powered by a pair of Bristol Siddeley Viper turbojets located side-by-side in the extreme nose of the aeroplane exhausting via nozzles on the sides of the aeroplane on the centre of gravity. For vertical take-off the nozzles were vectored directly downward, and for transition into forward flight were vectored gradually aft. The first successful transition was accomplished in May 1958, and the aeroplane was later re-engined with General Electric J85 turbojets.

The company by this time was more concerned with vertical take-off than with high speed, and its next prototype in the X-series was the extraordinary X-22A, which was designed for evaluation of the tilting-duct concept in an airframe that might form the basis of a light transport. The rectangular fuselage accommodated at its rear a wide-chord wing fitted on its leading edges with two groups of two 1250-shp (932-kW) General Electric YT58-GE-8D turboshafts to drive four propellers (or fans) located inside annular ducts. These last were located at the tips of the wings and the short-span foreplane, and could be turned between the vertical (for vertical take-off and

ABOVE. One of many experimental prototypes produced by Bell from the mid-1940s onward, the X-22A was an experimental tilting-duct type with four ducted propellers that could be rotated between the vertical for vertical take-off and landing, to the horizontal for wingborne flight.

landing) and the horizontal (for forward flight). Change in the propeller pitch produced the thrust modulation for control, and was supplemented by movement of aileron in the slipstream of each duct. The first of two prototypes flew in March 1966, and the type proved highly successful in the development of the tilting-duct concept and its practical applications.

A different type of progress had already been achieved with the Bell XV-3, which was designed to evaluate the tilting-rotor type of convertiplane. In this, large-diameter propeller/rotor units located at the wingtips can be angled with their axes vertical to produce vertical lift before the aeroplane converts into wingborne forward flight as the units are angled forward until their axes are horizontal. The XV-3 was powered by a 450-hp (336-kW) Pratt & Whitney R-985 piston engine located in the rear fuselage to drive both propeller/rotor units via a system of gears and transmission shafts. The type first flew in August 1955, and achieved very useful results that ultimately paved the way for the Bell XV-15. This was produced to a design approximating that of a production type,

The Bell XV-3 was produced as a tilting-rotor convertiplane, and its success should help see the introduction of the Bell/Boeing V-22 Osprey as the world's first operational convertiplane.

and powered by two 1,800-shp (1342-kW) Allison T53 turboshafts first flew in 1976.

The other major American exponent of such aircraft has been Boeing Vertol, which in August 1957 flew the VZ-2A tilting-wing prototype. Such a design as affinities to the tilting-rotor concept, but for reduced engineering problems uses propeller/rotor units in a fixed installation on a wing that is arranged to tilt between the vertical and horizontal as required by the crew. The VZ-ZA was powered by an 860-shp (641-kW) Avco Lycoming YT53-L-1 turboshaft and completed many successful flights. The success of these two differing but similar approaches to a wingborne transport with vertical take-off capability led to the selection of Bell and Boeing to produce the type planned as the world's first operational tilt-rotor transport, the Bell/Boeing V-22 Osprey that made its initial flight in 1989.

Another contender in the same basic field was Curtiss-Wright, the first of whose X-19A prototypes flew in June 1964 after development from the X-100 that took to the air for the first time in March 1960 on the power of a single 825-shp (615-kW) Avco Lycoming YT53-L-1 turboshaft. Akin to the X-22A in its concept of using four swivelling lift/thrust units located at the 'corners' of the airframe, the X-19A was powered by two 2,200-shp (1640-kW) Avco Lycoming T55-L-5 turboshafts driving four propeller/rotor units.

Other prototypes in this VTOL arena were the Doak VZ/4DA convertiplane, the Fairchild VZ-5FA deflected-slipstream V/STOL aeroplane, the Hiller X-18 tilt-wing convertiplane, the Lockheed XV-4A Hummingbird augmented-jet ejector lift aeroplane, and the Ryan VZ-3RY Vertiplane deflected-slipstream V/STOL aeroplane.

The VZ-4DA first flew in February 1958 with an 840-shp (626-kW) YT53 turboshaft driving a tilting ducted fan unit at each wingtip. The VZ-5FA made its initial flight in November 1959, and was powered by a single 1024-shp (763-kW) YT58-GE-2 turboshaft driving four larger-diameter propellers along virtually the full span of the wing leading edges where their thrust could be deflected vertically downward by large-area VTOL flaps. The X-18 took to the air for the first time in November 1959, and in overall concept was similar to the VZ-2A although a larger machine configured as a transport type. Power was provided by two 5850-shp (4362-kW) Allison T40-A-14 turboshafts driving the two contra-rotating propeller/rotor units located one on each wing, and a 3400-lb (1542-kg) thrust J34 turbojet providing exhaust gases to a tail-mounted thrust diverter used for longitudinal control in vertical flight.

The XV-4A first flew in July 1962 on the power of two 3300-lb (1497-kg) thrust Pratt & Whitney JT12A-3 turbojets. These were used for both vertical and forward flight by means of a series of diverter valves in the jetpipes. The aeroplane functioned conventionally in forwad flight, but for vertical flight the flaps were used to divert the thrust of the engines through a pair of ejector ducts in the fuselage. In this mode the exhaust was directed down through 20 transverse rows of multiple nozzles into the ejector chambers, where they mixed with cold air drawn through the open top doors in the fuselage and so boosted direct thrust by about 40 per cent. The VZ-3RY was similar in concept to the VZ-5FA and first flew in December 1958 with a 1000-shp (746-kW) T53-L-1 turboshaft driving two propeller/rotor units.

Ryan also produced another two VTOL aircraft as the X-13 Vertijet and XV-5. The former was a 'tail-sitter' in the mould of the Convair XFY-1 and Lockheed XFV-1, though in this instance configured as a pure research type powered by a single 10,000-lb (4536-kg) Rolls-Royce Avon turbojet. The aeroplane first flew in conventional mode with temporary wheeled landing gear in December 1955, and in 'tail-sitter' mode during May 1956. The XV-5 type pioneered the 'fan-in-wing' concept and first flew in May 1964.

Power was provided by two 2658-lb (1205-kg) thrust General Electric J85-GE-5 turbojets, whose power could be divertd to operate two wing-mounted fans which provided direct lift as required.

The other main developer of high-speed research aircraft in the USA was Douglas. The company's initial essay in this field was the D-558-1 Skystreak which first flew in May 1947 at the beginning of a programme to investigate free-flight air load measurements of the type that were then unobtainable in wing tunnel tests. The D-558-1 was fitted with a pressure-recording system connected to 400 points on the aeroplane's surface, and powered by a 4,000-lb (1814-kg) Allison J35-A-23 turbojet yielded invaluable research data. The type was later re-engined with the 5000-lb (2268-kg) thrust J35-A-11, and secured two world air speed records during 1947. The same company's D-558-2 Skyrocket was in essence a swept-wing version of the straight-winged D-558-1 powered by a 3000-lb (1361-kg) thrust Westinghouse J34-WE-22 turbojet supplemented by a 6000-lb (2722-kg) thrust Reaction Motors XLR-8 rocket motor, and was used for investigation of the performance and handling characteristics of swept wings. The type first flew in February 1948, and achieved enormously important results, including the world's first flight at over Mach 2 in November 1953.

Considerably more was expected of the Douglas X-3, often known as the Stiletto because of its very long fuselage with its finely tapered forward section, and its straight but diminutive flying surfaces. The type was schemed for the investigation of the thermodynamic problems of flight at up to Mach 3, the performance of turbojet engines at high Mach numbers, and the characteristics of double-wedge flying surfaces at high speed. The X-3 first flew in October 1952, but failed to live up to its potential because the powerplant of two 4200-lb (1905-kg) afterburning thrust Westinghouse J34-WE-7 turbojets was wholly inadequate.

The USA also put considerable effort into the creation of lifting-body vehicles as the design precursors of manned re-entry vehicles. These lifting-body vehicles were intended to prove the viability of wingless flying machines that could re-enter the atmosphere at hypersonic speed after orbital flight and fly back to their bases. The two main protagonists of such vehicles were Martin Marietta and Northrop, the former with the X-24, and the latter with the M2-F2 and HL-10. Developed from the unmanned X-23A, the X-24A had bluff contours and was powered by an 8000-lb (3629-lb) thrust Thiokol XLR-11 rocket engine. Air-launched from a Boeing B-52 motherplane, the X-24A first flew in powered form during March 1970 and revealed perfectly adequate handling as well as maximum speed and altitude of Mach 1.62 and 71,410 ft (32,390 m) respectively. The aeroplane was then reduced to its structural core and rebuilt as the X-24B with new lines including a sharp nose and triple rather than double vertical tail surfaces. In this form the type first flew in August 1973, and again proved highly successful.

Based on the wooden M2-F1 glider, the M2-F2 was similar to the X-24A in overall design, and was powered by the same type of rocket. Trials began in July 1956, and after an initial series of gliding flights had been achieved after air drop from a B-52 motherplane, the M2-F2 made its first powered flight in 1967. The machine was later rebuilt as the M2-F3 with triple rather than double vertical tail surfaces, and before the end of the test programme in 1972 yielded a mass of invaluable data. the HL-10 was similar to the M2-F2 in most respects other than the camber of its D-section lifting body: in the M2-F2 the flat and curved surfaces were on the top and bottom respectively, but on the HL-10 these positions were reversed. The type was again powered by the XLR-11 rocket engine, and after a series of gliding flights from December 1966 was gradually developed into the powered mode.

RIGHT. The Rockwell XFV-12A was a demonstrator for the concept of STOVL capability using the thrust augmented wing concept, and though many problems were encountered, promising results were finally achieved.

ROCKWELL HiMAT

In the 1970s there was considerable debate within aeronautical circles about the limits of the manoeuvrability potentially attainable with existing structural and propulsion technologies. To explore the limits of this manoeuvre envelope, Rockwell was contracted in a NASA and US Air Force programme to produce two examples of the HiMAT (Highly Manoeuvrable Aircraft Technology) design. This was a remotely piloted research vehicle, such a configuration being adopted for the twin reasons that the flight vehicle could be smaller (and therefore cheaper to build and operate) and also capable of sustained manoeuvre loads beyond those sustainable by pilots.

The HiMAT design was thus stressed for 12-g loads, and developed as a modular type in which any one or more basic component could be altered for the evaluation of factors such as different planforms, altered control surface relationships, various types of thrust-vectoring engine nozzles and modified aerofoils of the supercritical type offering high lift and low drag in the transonic speed range. The core of the vehicle was the fuselage complete with ventral air inlet and two aft-mounted structural booms, and extendable tricycle landing skids were incorporated. In its baseline configuration the HiMAT vehicle was completed with modestly swept aft-mounted wings, outward-canted vertical tail surfaces located on short beams alongside the engine jetpipe and an amazingly large set of flapped and sharply dihedralled canard foreplanes with about the same area as the main wings. About 90% of the skinning was made of graphite fibre composite. The HiMAT was capable of conventional take-off, but for extended mission time was generally carried to an altitude of 45,000 ft (13,715 m) under the starboard wing of a Boeing NB-52 Stratofortress motherplane before air launch. Control was exercised by the pilot in a ground station, though back-up control capability was provided by another pilot carried in the Lockheed TF-104G Starfighter chase plane.

The first HiMAT was delivered in June 1978, and the two vehicles made their first flights in July 1979 and July 1981 respectively. The flight test programme investigated a large number of control

The Rockwell HiMAT was a remotely piloted vehicle used for investigation of manoeuvre at very high *g* loads in a number of aerodynamic configurations.

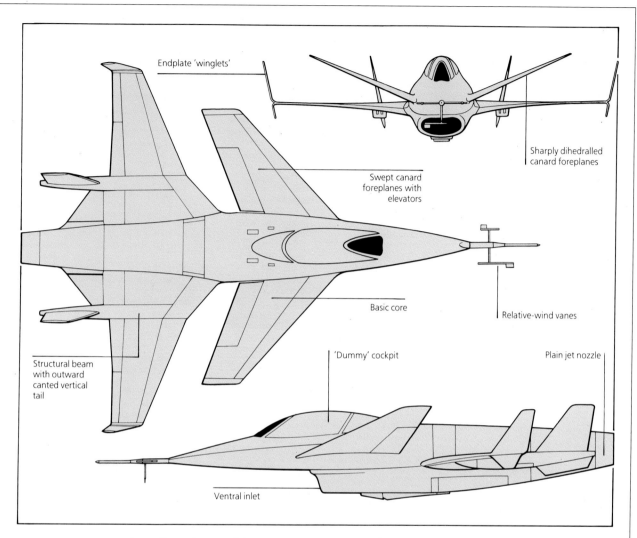

- Endplate 'winglets'
- Sharply dihedralled canard foreplanes
- Swept canard foreplanes with elevators
- Basic core
- Relative-wind vanes
- Structural beam with outward canted vertical tail
- 'Dummy' cockpit
- Plain jet nozzle
- Ventral inlet

configurations and dynamics, the performance of composite structures (including torsional stiffness as part of the investigation of aeroelastically tailored flying surfaces), and the interaction of close-coupled canard foreplanes, wing, winglets and fins. The HiMAT proved exceptionally manoeuvrable, providing approximately twice the capability of current fighters such as the General Dynamics F-16 Fighting Falcon and McDonnell Douglas F-15 Eagle. In general the flight tests fully confirmed the expectations of the design team, yet at the end of the 1983 flight test season the HiMAT programme was terminated by a cut-off of funding. This was particularly unfortunate, for the HiMAT had yet to be flown in some of the more ambitious configurations made possible by the vehicle's modular design and construction.

SPECIFICATION
Rockwell HiMAT

Type: remotely piloted high-agility research vehicle
Accommodation: none
Electronics and operational equipment: control system with TV, telemetry and radar links, and a suite of 164 research instruments
Powerplant: one 5000-lb (2268-kg) afterburning thrust General Electric J85-GE-21 turbojet
Performance: maximum speed 1060 mph (1710 km/h) or Mach 1.6 at high altitude; average endurance 30 minutes
Weights: empty 2645 lb (1200 kg); maximum take-off 3370 lb (1528 kg)
Dimensions: span 15 ft 7.25 in (4.755 m); length 22 ft 6 in (6.86 m) including probe; height 4 ft 3.6 in (1.31 m); wing area not revealed

INDEX

Page numbers in *italic* refer to captions

A

Adour *see* Turboméca
Advanced Tactical Fighter: 51
aero-engines *see under* manufacturer's name; jet engines, introduction of; RD series; rocket propulsion
Aerocentre
  NC. 1071: *179*
  NC. 1080: *179*
  NC. 270: *126*
  NC. 271: *126*
aeroelastic divergence: 62
afterburner (reheat): 74
AFTI *see* General Dynamics
Agile Falcon *see* General Dynamics
AIM-132 ASRAAM: *14*
AIM-54 (A/B/C variants) Phoenix: 55, 60
AIM-7 Sparrow: 13 *13*
AIM-9 (B/C/D/E/G/H/L/M/N/R variants) Sidewinder: 13, *14*
air-launched aircraft: 108, 131, 148, 197
Airbus
  A300-600 and A310: *87*
Allison
  J35-A-35: 76
  turbojets: 102, 105, 129, 197, 202
  turboprops: 131, 138
  turboshafts: 201
aluminium alloys: 10, 61
aluminium-lithium alloys: 68
America, USS: 60
Anglo-French Variable Geometry (AFVG) project: 181
Apache *see* McDonnell Douglas
Armstrong Siddeley
  Screamer: 178
  Snarler: 176, 178
  Spectre: 178
  turbojets: 178, 180-1, 199
Arsenal
  VG.70: *179*
  VG.90: *179*
Australia: 175
Avco Lycoming
  ALF 502-5: *88*
  turboshaft: 201
Avon *see* Rolls Royce
Avro
  Type 698: *125*
  Type 707: 124, *125*
  Type 720: 178
  Type 728: 178
  Vulcan: 124, *127*

B

Balzac *see* Dassault
Banshee *see* McDonnell
Bell
  AH-1 Huey Cobra: 68
  UH-1 Huey: 68
  X-14: 199
  X-1: *196*, 196, 197
  X-22A: 199-201, *199*
  X-2: 197
  X-5: 140, 197, *198*, 199
  XP-83: 132
  XV-15: 200-1
  XV-3: 200
Bell/Boeing
  V-22 Osprey: 200-1
Black Widow *see* Northrop
Blackbird *see* Lockheed
Blackhawk *see* Curtiss-Wright
Blackjack *see* Tupolev
BMW 003A turbojet: 158
Boeing: 14, 26, 40, 49, 55, 68, 75, 78, 87, 94-7, 100, 104-6, 108-9, 111, 116, 118, 130, 197, 202
Boeing Vertol
  VZ-2A: 201
bombers
  French: 125-6
  maritime attack: 119, 121-2
  Soviet: 116-23
  strategic: 101-5, 116-19, 128
  supersonic: 106-9, 119-22
  UK: 124-5
  US: 100-15
  variable geometry: 122-3
  see also Stealth
Bristol Siddeley turbojets: 186, *189*, 191, 199
Bristol Aerospace
  BAe 146: *88*
  EAP research fighter: *80*
  Harrier: 36, 41-9, *42*, 44, 46, 90
  Harrier GR Mk3: *42*, 44, 70
  Harrier GR Mk5: *43*
  Hawk trainer: *83*
  McDonnell Douglas/British Aerospace Harrier II: *41*, 48, *189*
  Sea Harrier: 48
  Sea Harrier FRS Mk1: *42*
bypass ratio: 82, 86

C

Canada: 174
canard configuration: 49-51, *51*, 57, 109, 146, 161, 194-5, 205
Canberra *see* English Electric
cannon: 13
carbon-fibre-reinforced plastic (CFRP): 65
Carter, Jimmy, President: 32, 110
Cayuse *see* Hughes
CCV *see* control-configured vehicles
Central Aerodynamics and Hydrodynamics Institute (USSR): 160, 166
CFRP *see* carbon-fibre-reinforced plastic
chines *see* LERXes
Chinook *see* Boeing
civilian aircraft: 108, 178
Cobra *see* Bell
Coleopter *see* SNECMA
Comet *see* de Havilland
Common Strategic Rotary Launcher: 55
composite powerplants: 133-4, 176, 178
  see also fibre-reinforced composite
Concorde: *87*
conformal weapon carriage: 52-7
Consolidated
  B-24: 130
  Model 37: 100, *102*
  XB-46: *104*, 105
Constellation *see* Lockheed
Constellation, USS: 61
control-configured vehicles (CCVs): 16-21, 51
Convair
  B-36: 100-3, *100*, 105, 116, 130-2
  B-58 Hustler: 106
  development: 104
  F-102 Delta Dagger: 55, 129, 145
  F-106 Delta Dart: *17*, 55, 145
  Model 7: 128-9
  Model 8: 129
  XB-36: 100-2, *100*
  XF2Y Sea Dart: 140
  XFY-1 Pogo: 138-40, *139*
  YB-36G: 103
  YB-60: *102-3*
convergent/divergent nozzle: 86
convertiplane: 199-201
core aircraft: 57-61
Cougar *see* Grumman
crescent wings: 124
cruise missiles: 104, 109, 122
Crusader *see* Vought
Curtiss-Wright
  prototypes: 132, 201
  rocket engine: 197
  X-100: 201
  X-19A: 201
  XP-87 Nighthawk: 132-3
  XP-87A Blackhawk: 132-3
Cutlass *see* Vought

D

Dassault: 179, 181, 186
  4A Mystere: 179-80
  B1 Super Mystere: 180
  Balzac: 186
  Ouragan: *179*
Dassault-Breguet 22, 23, 28, 29, 30, 48, 133, 180-5
de Havilland
  Comet: 178
  development: 176
  Mosquito: 63
  Sea Vixen: 176
  Sprite: 178
  turbojets: 133, 178
  Vampire: 174, 176
  Venom: 176
Delta Dagger *see* Convair
Delta Dart *see* Convair
Demon *see* McDonnell
Derwent *see* Rolls Royce
digital flight-control system *see* fly-by-wire control
direct lift: 163, 202
direct-control modes: 51-2
Doak: 201
  VZ/4DA convertiplane: 201
Dornier
  Do 31E: *189*, 191
Douglas: 105, 106, 136, 138, 143-4
Draken *see* Saab
Dutch rolling: 51

E

Eagle *see* McDonnell Douglas
Eagle (AAM-N-10) missiles: 144
Emerson Electric: 108
engines (powerplant): 74-97
  radial: 75
English Electric
  Canberra: 106, *126*
  Lightning F Mk6: *74*
  P.17A: 178
  P.1: 177
  P.1 Lightning: *177*
Enhanced Fighter Manoeuvrability programme: 148
Entwicklungsring Sud Arbeitsgemeinschaft (ESA): 191
ESA
  VJ-101C: 191
European Fighter Aircraft: 51
experimental aircraft *see* research aircraft

F

Fairchild
  VZ-5FA: 201
Fairchild/Republic
  A-10A Thunderbolt II: 79
Fairey
  F.155T: 178
Falcon *see* General Dynamics
Falcon
  GAR-1 missiles: 145
  GAR-3 missiles: 145
Farmer *see* Mikoyan-Gurevich
FAST packs: *17*
Fencer *see* Sukhoi
fibre-reinforced composite (FRC): 62, 63, 64, 68, 84
fighter-bombers: 146, 162, 164-7
fighters: airlaunched: 131-2
  all-weather: 157-9, 166, 172-3, 175
  American: 128-53
  direct lift: 163, 186
  English: 174-8, 189
  escort: 130-1
  French: 178-81, *182*, *183*, *184*, *185*
  German: 128
  naval: 133-45
  night: 132, 136
  rocket-powered: 176, 178-9, 181
  role: 128, 130
  Soviet: 154-73
  supersonic: 136, 158, 172
  US F-series: 149-52
  variable geometry: 163-4, 170, 181
  and VTOL: 146, 173, 186-91
Fighting Falcon *see* General Dynamics
Fireball *see* Ryan
Fishbed-K *see* Mikoyan-Gurevich
Fitter *see* Sukhoi
Flogger *see* Mikoyan-Gurevich
fly-by-wire control (digital flight-control system): 9, 21-2, 52
Flying Bedstead: 188
flying boom: *87*
Flying Fortress *see* Boeing
flying wing: *101*, 102-3, 114, 122-3, 136
Ford Aerospace: 14
Forger: *see* Yakovlev
Foxbat *see* Mikoyan-Gurevich
France
  Anglo-French collaboration: 181
  bombers: 125-6
  fighters: 178-85
  naval: 179
  variable geometry aircraft: 181
  and VTOL: 181, 186
FRC *see* fibre-reinforced composite
Freedom Fighter *see* Northrop
Frogfoot *see* Sukhoi
Fulcrum *see* Mikoyan-Gurevich
Fury *see* Hawker and North American
fuselage chines *see* LERXes

G

General Dynamics
  401: 82
  F-111: 22, 25, 29, 34, 34, 108, 122, 142, 149, 163
  F-111A: 24, 25
  F-111B: 55, 122, 142-3
  F-111D: 25
  F-16 Agile Falcon: 20
  F-16 Fighting Falcon: 14-16, *17*, 20, 21, *21*, 22, 22, 48, 48, 52, 54, 55, 82, 83, 84, 152, 205
  F-16/AFTI: 20, 48, *48*, 52
  F-16B: 84
  F16C: 20, 52
  F-16D: 52
  F-16XL: 52-5, 84, 150-1, 152-3
  FB-111A: *109*
  FB-111H: *110-11*
  Model 401: 152
  YF-16: *14*, 19, 52, 146, 152
General Electric
  F101-GE-100: 97
  F101-GE-102: 97
  F404: 64, 82, 84
  J101: 82
  J79-GE-17: 76
  TF34: 79
  turbofans: 109, 115, 130, 146, 186
  turbojets: 105, 107, 132-3, 144, 146, 199, 202, 205
  turboprops: 133
  turboshafts: 201
  UDF engine: *87*
geodetic construction: 62
Gerfaut *see* Nord
Germany: 116, 130
Germany
  West and VTOL: 191
  World War II data: 103-4, 118, 128, 136, 154, 197
glass-reinforced plastic (GRP): 62
gliders: 109, *126*, 202
Gloster
  CXP-1000: 175
  G.42 (E.1/44): 174-5
  GA.5 Javelin: 175
  jet fighters and: 174-5
  Meteor: 74, 156, *174*, 175
Goblin *see* McDonnell
Griffon *see* Nord
Gripen *see* Saab
GRP *see* glass-reinforced plastic
Grumman: 30, 34, 55-60, 64-8, 107, 135-6, 140-3, 152
guns: 13

H

Halo *see* Mil
Handley Page
  H.P.88: 124
  Victor: *124*
Harpoon *see* McDonnell Douglas
Harrier *see* Hawker, British Aerospace and McDonnell Douglas
Havoc *see* Douglas
Hawk *see* British Aerospace and McDonnell Douglas
Havoc *see* Douglas
Hawk *see* British Aerospace
Hawker: 90, 156, 174-6, 178, 187-9, 191
head-up displays (HUD): 52
helicopters: 68
Hiller
  X-18: 201

HiMAT (Highly
    Manoeuvrable
    Aircraft Technology)
    see Rockwell
Hind see Mil
Hornet see McDonnell
    Douglas
hose-and-drogue system:
    87
HOTAS system: 21-2
HUDs see head-up
    displays
Huey see Bell
Hughes
    AIM-54 (A/B/C
        variants) Phoenix: 55,
        60
    air-to-air missiles: 55,
        143
    APG-70: 17
    OH-6 Cayuse: 68
Hummingbird see
    Lockheed
Hunter see Hawker
Hustler see Convair
hydro-skis: 140

I

I designations see
    Mikoyan-Gurevich
IAI see Israel Aircraft
    Industries
ICBMs see missiles
Il designations
    Il-28: 117-21, 126
    Il-30 119, 120-1
    Il-46: 121
    Il-54 120-1, 120
Ilyushin design bureau:
    120-1
in-flight refuelling: 87
infra-red tracking
    systems: 112-122
inlets: 86-7, 91-3, 97
Invader see Douglas
Israel Aircraft Industries
    (IAI)
    Kfir: 49
    Kfir-C2: 48, 48
    Kfir-TC2: 48
    Lavi: 20, 93, 94, 191,
        192-3

J

Jaguar see Grumman
jamming see electronic
    countermeasures
Javelin see Gloster
jet engines, introduction
    of: 102-4, 116-17, 125,
    128, 154, 158, 171
Jumo see Junkers
Junkers
    Ju 287: 62
    Jumbo turbojet: 154-5,
        171, 179; see also
        RD series
    Jumo 109-004: 76
    Jumo 109-004A: 76
    Jumo 109-004B-1: 76
    Jumo 109-004B-4: 76
    Jumo 109-004D-4: 76
    Jumo 109-009: 76

K

Kestrel see Hawker

Kfir see Israel Aircraft
    Industries
Klimov turbojets: 118, 157,
    162, 172
Koliesov turbojets: 90,
    120, 163
Korean War: 131, 136, 177
Krüger flap: 40, 40

L

La designations see
    Lavochkin
LANTIRN system: 17
Lavi see Israel Aircraft
    Industries
Lavochkin: 154-9, 166
leading-edge vortex
    generator see vortex
    generator
LERXes (fuselage chines):
    51
Light-Weight Fighter
    Competition (1974):
    152
Lightning see English
    Electric
Lippisch, Dr Alexander:
    128
Lockheed
    A-11: 149, 152
    F-104 Starfighter: 145
    F-104 Starfighter: 26,
        26, 28
    F-19: 93
    F-80 Shooting Star:
        130
    F-90: 129
    F-94 Starfighter: 132
    F-94 Starfire: 132, 133
    L-1011-500 TriStar: 52,
        53
    L-749 Constellation: 75
    Model 153: 130
    S-3: 79
    SR-71 Blackbird: 144-
        5, 149
    SR-71 and SR-71A
        Blackbird: 10-11, 11,
        74, 92, 97, 144, 145, 149
    U-2: 10
    XV-4A Hummingbird:
        20
    XVF-1: 138, 140
    YF-12A: 149, 152
losses: 103, 166, 176, 178,
    179, 181, 197
LR-11 rocket: 128, 130
Lynx see Westland
Lyul'ka turbojets: 90, 120-
    1, 157-8, 162-3, 166,
    172

M

McDonnell
    development: 131
    F2H Banshee: 135-6
    F3H Demon: 136, 137
    F-101 Voodoo: 131,
        143, 145
    FH-1 Phantom: 133
    Model 27: 131
    MX-472: 131
    XF-85 Goblin: 131-2,
        198
    XF-88: 130-1
McDonnell Douglas: 8, 9,
    13, 14, 16, 17, 19, 37,
    41-3, 46, 51, 68, 69,

70, 78, 84, 91-3, 144-6
    152, 205
    see also British
        Aerospace,
        Douglas
McDonnell, Douglas/
    Northrop
    F/A-18A Hornet: 152
Marauder see Martin
Marboré see Turboméca
maritime attack role: 118,
    121-2
Martin
    B-26 Marauder: 104
    B-57: 106
    XB-48: 105
    XB-51: 107
Martin Marietta
    X-23A: 202
    X-24 glider: 202
MAW (mission adaptive
    wings): 22-30, 65, 68
MBB (Messerschmidt-
    Bolkow-Blohm): 148
Messerschmitt
    Bf 110: 76
    Me 163: 128
    Me 262: 74, 76
    Me 262A: 76
    P.1011: 140
    P.1101: 197
Meteor see Gloster
MiG designations see
    Mikoyan-Gurevich
Mikoyan-Gurevich: 12, 13,
    28, 35, 74, 91, 92, 112
    119, 131, 134, 154, 156,
    158, 161, 162, 164, 166,
    171-2
Mil
    Mi-24 Hind: 68
    Mi-26 Halo: 68
Miles
    M.52: 174
Minsk: 90
Mirage see Dassault-
    Breguet
Missileer see Douglas
missiles: 55, 68, 78
    air-to-air: 14, 55
    airborne: 101, 114, 121-
        2, 124, 128, 136,
        142-6, 152, 158,
        161
    anti-aircraft: 108
    cruise: 78, 104, 109,
        122
    ICBM: 109, 121, 178
    surface-to-air: 12, 14
    see also AIM
Mitchell see North
    American
Mixmaster see Douglas
Mosquito see de
    Havilland
Multi-Role Combat
    Aircraft (MRCA): 181
Myasishchyev
    DBV-202: 116
    design bureau: 116,
        118-20
    M-4: 117-20
    M-50: 119-21
    M-52: 120-1
Mystere see Dassault

N

NASA (National
    Aeronautics and Space

Administration): 140, 204
naval aircraft
    anti-submarine
        aircraft: 119
    flight-deck restrictions:
        136, 138, 140
    French Navy: 179
    Royal Navy: 175, 177
    Soviet Navy: 173
    US Navy: 133-45
    variable geometry:
        140-3
    VTOL: 138-40
    waterplane fighter:
        140
Naval Ordnance Test
    Station (Naval
    Weapons Center): 14
Nene see Rolls Royce
Nighthawk see Curtiss-
    Wright
Nord
    2200: 179
    N.1402 Gerfaut: 180-1
    N.1500 Griffon: 181
    ramjet: 181
North American
    A-5 Vigilante: 164
    B-25 Mitchell: 104
    B-45 Tornado: 105,
        126
    B-70 Valkyrie: 164
    F-100 Super Sabre:
        28, 28, 142, 145-6
    F-86 Sabre: 91, 93,
        128, 130, 133, 136,
        145
    FH-2 Fury: 133
    FJ-1 Fury: 128, 133,
        133, 135
    FJ-2 Fury: 136
    FJ-4 Fury: 138
    P-51: 130
    XB-70 Valkyrie: 106-9,
        163-4
    XFJ-1: 133
    YF-108 Rapier: 146
North American/
    Rockwell: 108
Northrop: 14, 16, 30, 64,
    102-3, 111, 131, 132,
    147, 149, 152, 202
Northrop, Jack: 30
nozzles: 86-91, 97
nuclear weapons: 100-2,
    104, 109, 111, 113-16,
    121-2, 124-5, 128, 145

O

Olympus see Rolls-Royce
Osprey see Bell/Boeing
Ouragan see Dassault

P

P.177 rocket engine: 178
Panavia
    Tornado: 29, 30, 34,
        37-40, 38, 76, 87,
        88, 105, 126, 181
    Tornado ADV: 54, 80
    Tornado GR Mk1: 38,
        88
    Tornado IDS: 39, 54
Panther see Grumman
parasite fighters see air-
    launched aircraft
Pegasus see Rolls-Royce

Petlyakov
    Pe-8: 116
Phantom see McDonnell
    Douglas
Philco Ford: 14
Phoenix see Hughes
Pirate see Vought
piston engines: 101-3- 117,
    133-4, 158, 171, 200
plenum-chamber
    burning: 91, 91
Pogo see Convair
Power Jets (Whittle)
    turbojet: 174
powerplant see engines
Pratt & Whitney
    F100: 82, 83, 84, 93
    JT11D: 92
    P557-P-43: 94
    JT9D-7RA: 87
    piston engines: 102,
        133-4
    PW1120: 93
    TF30-P-412: 88
    turbofans: 143
    turbojets: 103, 129,
        131, 136, 146, 149,
        201
probe-and-drogue
    system: 87
Pterodactyl see Westland

R

radar: 17, 93-4, 122, 125,
    128, 132, 136, 142,
    145, 157-8, 162, 167,
    173, 180
    avoidance: 111-14
    cross section (RCS):
        93-4
    ground defence: 108
    radar-absorbent
        material (RAM):
        93-4
radial engines see
    engines
Rafale see Dessault-
    Breguet
RAM see radar
Rapier see North
    American
Raytheon: 14
RCS see radar
RD series turbojets: 154-6,
    158-9, 164, 166, 171
Reaction Motors rocket
    engines: 196, 202
Reagan, Ronald,
    President: 110-11
reconnaissance aircraft:
    102, 118-19, 121, 131,
    145, 149, 178
refuelling see in-flight
    refuelling
reheat see afterburner
Republic
    F-105 Thunderchief:
        143, 145
    F-84 Thunderjet: 129
    F84F Thunderstreak:
        129
    P-47: 130
    XF-91 Thunderceptor:
        129-30
research aircraft: 126,
    155, 161, 170, 196-205
    high-altitude: 108-9
    high-speed: 108-9,
        129, 196-7, 202

lifting-body: 202
    VTOL: 199-202
    wing research: 124,
        128-9, 134-5,
        197-9, 198
rocket propulsion: 108,
    126, 128, 130, 145,
    161, 175-6, 178-9,
    196-7, 202
rockets: 78, 128
Rockwell
    B-1: 30, 34
    B-1A 109, 110, 111
    B-1A: 32, 33, 97, 97
    B-1B: 33, 94, 97, 97,
        104, 111-12, 114
    development: 108-9
    HiMAT: 57, 57, 148,
        204-5
    SR-71
        S-1010B-suit: 10
    X-15: 108-9
    XFV-12A: 203
Rockwell/MBB
    TKF-90: 148
    X-31A: 148-9, 148-9
Rolls-Royce
    Avon: 93, 125, 176, 201
    Derwent: 117-18, 155-
        6, 171
    Nene: 117, 126, 155,
        175-6, 179
    Olympus 593 (with
        SNECMA): 87
    Pegasus: 82, 83, 90, 91
    RB211: 87
    RB211-535C: 87, 88
    RB series: 186, 188,
        191
    Tay: 176, 180
    see also Turboméca
Ryan
    F-109: 146
    FR-1 Fireball: 133
    VZ-3RY Vertiplane:
        201
    X-13 Vertijet: 146, 201
    XV-5: 201-2

S

S designations see Sukhoi
Saab
    35 Draken: 36, 37
    37 Viggen: 36, 37, 49,
        51, 194-5
    AJ37 Viggen: 37
Sabre see North
    American
Saunders-Roe
    SR.53: 178
SCECMA
    Atar turbofans: 180-1
    TF-104 turbofans: 186
Scorpion see Northrop
Screamer see Armstrong
    Siddeley
Sea Dart see Convair
Sea Fury see Hawker
Sea Harrier see British
    Aerospace
Sea Hawk see Hawker
Sea King see Sikorsky/
    Westland
Sea Stallion see Sikorsky
Sea Vixen see de
    Havilland
Seahawk see Sikorsky
SEPECAT Jaguar: 181

SEPR rocket engines: 181-2
Shooting Star see Lockheed
Short
  SA.4 Sperrin: 124-5
  SC.1: 188
Shvetsov engine bureau: 166
Sidewinder see AIM-9
Sikorsky
  CH-53 Sea Stallion: 68
  Seahawk: 68
Sikorsky/Westland
  Sea King: 68
Skylancer see Douglas
Skynight see Douglas
Skyray see Douglas
Skyrocket see Douglas
Skystreak see Douglas
Snarler see Armstrong Siddeley
SNECMA see Rolls-Royce
SNECMA
  C.450 Coleoptere: 190, 191
South-East Asia War see Vietnam War
space
  near-space flight: 108
  re-entry: 109, 202
  satellite surveillance: 113, 123
Space Shuttle: 10, 109
Sparrow see AIM-7
Spectre see Armstrong Siddeley
Sperrin see Short
Sprite see de Havilland
Stalin, Joseph: 116
Starfighter see Lockheed
Stealth see Northrop
  stealth technology: 93-7
STOL: 34-7
STOVL: 41-9, 44, 90-1, 90
STOVL capability: 203
Strategic Air Command (SAC) requirements: 101-2, 104, 109, 130
Stratofortress see Boeing
Stratojet see Boeing
Strike Eagle see McDonnell Douglas
Su designations see Sukhoi
Sud-Est
  development: 181
  SE.212 Durandal: 181
  SE.2410 Grognard: 179
  SE.5000 Bardoudeur: 181

Sud-Ouest
  SO.4000: 126, 127
  SO.4050 Vautour: 126, 179
  SO.6020 Espadon: 179
  SO.6025 Espadon: 179, 179
  SO.6025 Espadon: 179
  SO.9000 Trident: *181*
  SO.M-1: 126
  SO.M-2: 126
Sukhoi
  Aircraft P: 166
  Aircraft R: 166
  design bureau: 154, 164-71
  P-1: 170
  Pt-9: 167
  S-1: 166-7
  S-221: 170
  Su-11: 166-7
  Su-13: 166
  Su-15: 166, 170
  Su-17 Fitter-C: 33, 166, 170
  Su-20: 170
  Su-21: 170
  Su-22: 162, 170
  Su-24 Fencer: 30, 34, 122-3
  Su-25 Frogfoot: 167, 168-9, 170-1
  Su-27: 122, 170-1, 173
  Su-71G: 170
  Su-7: 167, 170
  Su-7BMK Fitter-A: 33
  Su-9: 162, 164, 166-7
  T-37: 170
  T-3: 167
  T-49: 167
Super Mystere see Dassault
Super Sabre see North American
supercritical wing section: 29-30
Superfortress see Boeing
Supermarine
  Attacker: 176-7
  fighter development: 176-7
  Type 535: 177
  Type 510: 177
  Type 541 Swift: 177
  Type 545: 177
supersonic flight: 106-9, 119-22, 136, 141-2, 144-5, 158, 166-7, 172, 174, 177-8, 191, 196-7
SW-1: 14

Swift see Supermarine
swing-wing aircraft see variable geometry

T

Tay see Rolls Royce
Teledyne see McDonnell Douglas
Tempest see Hawker
thickness-chord ratio: 26-8, 51
Thiokol rocket engine: 202
three-poster propulsion: 91
Thrust Measuring Rig (Flying Bedstead): 188
thrust-reversal system: 87
thrust-to-weight ratio: 91
thrust-vectoring: 48-9, 91
Thunderbolt see Fairchild/Republic
Thunderceptor see Republic
Thunderchief see Republic
Thunderjet see Republic
Thunderstreak see Republic
Tiger see Grumman
tilting ducts: 199-201
tilting wings and rotors: 200-1
Tomcat see Grumman
Tornado see North American; Panavia and North American
transonic tails: 51-2
TriStar see Lockheed
Tu designations see Tupolev
Tumanskii
  R-31: 92
  turbojets: 160-1, 164, 173
Tupolev: 30, 116-19, 120, 121-3, 161, 166
Turbo-Union
  RB199: 76, 82, 88
  RB199 Mk104D: 80
turbofan: 76-87, 87
turbojet: 74-6
Turboméca: 180-1
  Marboré II: 78
  Rolls-Royce Adour: 82, 83
turboprop: 86, 87
turboshaft: 68
turbulator see vortex generator

U

U-2 (spyplane) see Lockheed
unducted fan (UDF): 86
Union of Soviet Socialist Republics
  air defence of: 106, 108-9, 112, 130
  bombers: 116-23
  fighters: 154-73
  research aircraft: 202
  stealth aircraft: 113, 123
  supersonic aircraft: 119-22, 156, 158
  VTOL: 173
Union of Soviet Socialist Republics, see also Lavochkin, Mikoyan-Gurevich, Myasishchyev, Sukhoi, Tupolev, Yakovlev
United Kingdom
  ballistic missiles: 178
  fighters: 174-8
  rocket propulsion: 176
  strategic bombers: 124-5
  turbine power: 174
  VTOL: 186, 188-9
United States of America
  bombers: 100-15
  research aircraft: 108
  Stealth programme: 109-15
  strategic bombing development 100
  fighters: 128-53
  air-launched: 131-2
  designation rationalisation: 146, 149
  lightweight: 151-2
  naval: 133-45
  night: 132-3
  research: 196-202
  high-speed flight: 196-7, 202
  lifting-body vehicles: 202
  VTOL: 199-200
  wing geometry: 197, 198, 199
  rocket-powered: 128, 145
  VTOL: 138-40, 139, 146, 199-202

US Marine Corps: 152
US Navy: 133-45
USAAF: 100, 103-5, 132
USAF: 103-6, 132, 145, 204

V

Valiant see Vickers
Valkyrie see Boeing and North American
Vampire see de Havilland
variable geometry (swing wing): 30-4, 32, 34, 37-40
  inlets: 93
variable-geometry aircraft: 109, 122-3, 140-3, 163-4, 197, 198, 199
variable-incidence wings: 105, 122, 130, 141-2
Venom see de Havilland
Vertijet see Ryan
Vertiplane see Ryan
VFW-Fokker
  VAK-191B: 191
Vickers
  G.4/31: 62
  M.1/30: 62
  Valiant: 124-5
Victor see Handley Page
Vietnam War: 12-14, 21, 152
VIFFing: 48-9
Viggen see Saab
Vigilante see North American
Viper see Armstrong Siddeley
Voodoo see McDonnell Douglas
vortex generator (turbulator): 51
Vought
  F5U: 134
  F6U-1 Pirate: *134, 135*
  F7U Cutlass: *136, 136*
  F8U Crusader III: *141*, 142
  V-173: 134, 135
  XF5U: *134, 135*
  XF6U Pirate: 134-5, 135
VSTOL (Very Short Take-off and Landing): 201
VTOL (Vertical Take-off and Landing) programmes: 41-9, 91, 186-95
  France: 186-8
  Germany: 191

UK: 188-9
USA: 138-40, 146, 199-202
USSR: 173
Vulcan see Avro

W

Wallis, Sir Barnes: 62
Warwick bomber: 62
waterplane: 140
Wellesley bomber: 62
Wellington bomber: 62
Westinghouse turbojets: 128, 130-3, 135, 140, 202
Westland
  Lynx: 68
  Pterodactyl IV: 140
Westland, see also Sikorsky
Whitcomb, Richard T.: 41
Windsor bomber: 62
winglets: 40, 41
wings
  adjustable: 17-21
  aft-swept (swept-back): 62, 65
  forward-swept: 62-8, 65
  mission-adaptive (MAW): 22-30, 65, 68
  variable-camber: 68
  see-also supercritical wing section and variable geometry
World War II: 55, 100-1, 103-5, 116, 128, 130-1, 136, 154, 158, 175, 178-9
Wright
  rocket engines: 145
  turbojets: 116, 145

X

XAAM-N-7: 14

Y

Yak designations see below
Yakovlev: 90, 120, 121, 154, 158, 171, 172, 173
Ye designations see Mikoyan-Gurevich
Yeager, Captain C E (Chuck): 196

PICTURE CREDITS AND ACKNOWLEDGEMENTS

All illustrations for this publication were supplied through Military Archive & Research Services, Lincs. All line artworks from Ravelin Limited. t =top; c =centre; b =bottom; l = left and r =right.

Boeing: p40, 75, 95, 97. | **British Aerospace:** p38, 39(t), 43(b), 44, 45, 46, 54(t), 63(t), 71(t), 74(t), 80(t), 81(t), 88(c), 89. **Commonwealth Aircraft Co:** p92(t).
Avions M Dassault-Brequet Aviation: p23, 29, 30, 49(b), 50.
Department of Defense: p12, 25, 34(tl), 55(t), 79, 90.
General Dynamics: p18, 20, 21, 22(t), 24(t), 48, 52, 54(t), 83(t), 84(br).
General Electric: p86(t&c). **Grumman Corp:** p55(b), 58, 59, 60, 64(b), 65, 88(b).
Imperial War Museum: p63(b). **Israel Aircraft Industries:** p49(t&c), 93(c&b), 94(b).
Robert Jackson: p100, 102/3, 105, 106/7, 116, 118/19, 120, 124/5, 126, 127(b), 128/9, 130/1, 132/3, 137, 138, 142/3, 154/5, 156, 166, 167(b), 170, 176/7, 179, 181, 198.
Lockheed Corp: p11, 26(t), 27, 53, 74(t), 77(b).
MARS: p14(b), 96(t), 101, 104, 105(b), 108/9, 110/11, 117(t), 127(t), 128(b), 134/5, 136, 139, 141, 144/5, 146/7, 148, 152/3, 157, 158/9, 160/1, 172, 174/5, 180, 182/3, 186/7, 188/9, 190/1, 194/5, 198(b), 199, 200. **MARS/Niska:** p34(b), 51. **Matra:** p22(b).
McDonnell Douglas: p8, 9, 17, 19, 41, 42, 43(t), 47, 64(t), 66, 67, 69, 70, 71(b), 78, 84(t&bl), 85, 92(b), 93(t). **Northrop Corp:** p31 **Panavia:** p39(b). **Pratt & Whitney:** p86(b).
Quadrant/Flight: p114/5, 122/3, 148/9, 167(t), 171, 173.
Rockwell International: p16, 28, 32, 33, 56, 57, 96(b).
Rolls-Royce: p80(b), 81(b), 82, 83(b), 87, 88(t), 91.
Royal Netherlands AF: p26(b). **Saab-Scania/Anderson:** p36, 37(c&b).
Salamander: p112/13, 150/1, 164/5, 168/9, 184/5, 192/3.
USAF: p10, 15, 24(c&b), 28(b), 76, 77(t), 94(t). **SIRPA:** p34(tr), 35.
US Navy: p13, 61. **Varo Ins:** p14(t). **Vickers Ltd:** p62